BLAZER AND ASHLAND OIL

PAUL G. BLAZER

Blazer
and
Ashland Oil

A Study in Management

by

JOSEPH L. MASSIE

UNIVERSITY OF KENTUCKY PRESS

*The publication of this book has been made possible partly
through a grant from the Margaret Voorhies Haggin Trust,
established in memory of her husband, James Ben Ali Haggin.*

To all of those in management
who give voluntarily of their time
in the interest of empirical research,
this book is dedicated.

PREFACE

SOME STUDENTS of management have been impressed by the universals of the administrative processes in the various areas of application and have ignored the specific setting of a managerial problem. Others prefer to qualify their statements by references to the surrounding facts of the specific situation. This study represents research based upon the second viewpoint. In this case, access to the details of the company history and to the executive organization has made possible a comprehensive study of the managerial processes as they evolved through time. These processes have been determined in part by events in the history of the company, in part by the institutional setting, and in part by personal characteristics of the individual executives.

Many works on management and organization have given greatest emphasis to the largest corporations, the civilian government, and the military forces. The managerial problems have been assumed to be the same—but simpler—for the smaller operators. This study raises the question of whether the approach to management by the chief executive of the smaller company might not be somewhat different from the approach indicated by writers who have studied the large firms.

The available sources of information made possible a comprehensive study of Ashland Oil & Refining Company, Ashland, Kentucky, a relatively small firm. Annual reports, audit reports, minutes of the board of directors, and the company magazine were available. In addition, letters, which from the company's inception had been written monthly by Paul G. Blazer, the chief executive, to the members of the board of directors, explained in detail the facts of the preceding month, current operational problems facing the company, decisions made and reasons for them, alternative courses of action, and opinions on the managerial processes. A complete file of these letters was located after an extended search and formed the basis for the research. In this way, it was possible to determine the detailed thoughts at the time of an action without depending upon memory.

These written sources provided the foundation for a series of personal interviews with those who had knowledge of the company's operations. Interviews were secured with a majority of the members of the board and all principal operating executives. Competitors, a union leader, and an investment counsel offered comments that helped to round out the complete picture of the company's organization and operations. However, by far the most important interviews were those with Blazer over a period of two and a half years. In these, the chief executive was able to elaborate on comments in the monthly letters and to add to his previous observations concerning his ideas on management. A tape recorder provided a means for more careful study of the facts and opinions expressed in many of the interviews. By replaying the recordings, it was possible to appraise the comments analytically with due regard for intensity of emphasis, inflection of voice, and other indications of the ultimate meaning of comments. By comparing the tape-recorded comments with the statements in the letters, it was possible to delineate the basic trend in Blazer's thinking. Rationalization could be detected more easily. Throughout most of the interviews a nondirective approach was relied upon, except in those instances in which certain specific information was desired for continuity.

By these methods it has been possible to approach the core of the thinking of one chief executive over a period of time and to organize his concepts of management in light of the factors as he saw them at the time and place of the specific decision. References to Blazer's written and verbal comments will be made extensively in order to support the principal statements.

In order to interpret managerial decisions and the fundamental managerial philosophy upon which they were based, it is mandatory to consider them as within the framework of the economic conditions and personal relationships which existed at the time. This study, therefore, examines the economic structure and functioning of the oil industry. The techniques of the business historian are required to establish the chronological development and the personal factors so necessary to a sound evaluation.

This study, undertaken initially for a doctoral dissertation at the University of Chicago, was made possible by the cooperation of Paul G. Blazer, former president and chairman of the board of directors of Ashland Oil & Refining Company. On September 15, 1952, he consented to make all company documents available which were necessary for an intensive study of the management. His interest in business research motivated him to devote his time to interviews. All requests for information were granted.

Company executives gave me their enthusiastic support. Arthur J. Points, controller, served as a valuable liaison between the operating personnel and myself. I am indebted especially to the following members of the board: E. L. McDonald, Lexington, Kentucky; the late James L. Martin, Chicago; Walton T. Davis, Louisville; and Robert D. Gordon and William C. Freeman, Chicago. Valuable information was provided in interviews with former owners of predecessor companies, including Earle M. Craig, Freedom, Pennsylvania, and W. H. Bennett, Buffalo, New York.

Earl Wallace, Lexington, Kentucky, former vice president of the Standard Oil Company (Ohio), furnished valuable information about Ashland Oil over its entire history as seen from the viewpoint of a competitor. J. Fred Miles, Louisville, granted interviews and access to early records in his possession.

Although the facts upon which this study is based have been made available from company records and other related sources, the members of the management of Ashland Oil & Refining Company do not subscribe to my interpretations and conclusions, which are the product of my own judgment and for which I accept personal responsibility.

J. L. M.

CONTENTS

EXHIBITS

EXHIBITS

ILLUSTRATIONS

1

THE SETTING FOR SUCCESS

Ashland Oil's Place in the Petroleum Industry

ALTHOUGH THE petroleum industry in the United States recognizes its beginning in 1859, when Edwin L. Drake drilled the first commercial oil well, near Titusville, Pennsylvania, its modern character dates from the early years of the twentieth century, when a series of events transformed its structure. The opening of the southwestern oilfields, announced spectacularly by the Spindletop gusher near Beaumont, Texas, in 1901, and the mass production of automobiles, initiated by the Model-T Ford in 1908, established a new supply-and-demand relationship for petroleum which resulted in a vast and rapid growth of the industry. In 1911 the old Standard Oil Company was broken up by a Supreme Court order, and the petroleum industry was thrown open to all comers.

Many came; and as the industry prospered, so did its members, large and small, whose enterprises were sound. But one of the smaller firms, Ashland Oil & Refining Company (and its predecessor, Ashland Refining Company), experienced a rate of growth that was exceptional even for an industry characterized by unusual expansion. This study presents an analysis and interpretation of the management of this Kentucky business by its chief executive, Paul G. Blazer, from its inception, through the period of growth, to a time of relatively large size. Its purpose is to analyze all of the important factors that have determined the management of that company and to organize the important components of the managerial philosophy of the chief executive into a unified body of thought. The focus is on the relationships of Blazer with the board of directors and with the operating officers and department heads.

The study of Blazer's managerial objectives and concepts necessarily is both a biography of the chief executive and a history of the company. This story covers a relatively short time. It begins in 1918, when J. Fred Miles, an Oklahoma oil promoter, organized a Kentucky crude oil producing company which six years later financed the formation of a small refining company. It ends in 1957 with the retirement of Paul G. Blazer, the chief executive of Ashland Oil & Refining Company, who in thirty-three years had built a company with a quarter of a billion dollars in sales.

The assets of the refining company grew from $250,000 in 1924, when it was organized with Blazer as manager, to $8,000,-000 by 1940; $24,000,000 by 1947; $67,000,000 by 1949; and $175,000,000 by 1956. Net income after taxes increased from $36,000 in 1924 to $13,500,000 in 1956. At no time during the entire history did the refining company end a year with a net loss. Its ratio of net income (after taxes) to owner's equity fluctuated between 6½ percent (in 1934) and 40 percent (in 1928 and 1948).

This growth has been accomplished by accretion and by merger of approximately fifty companies (the more important of which are listed in Exhibit 1). The most important accessions took place in 1930-1931 and 1948-1950. In the former period, the company acquired a refining company and a pipeline; in the latter, Allied Oil Company, Aetna Oil Company, Freedom-Valvoline Oil Company, and Frontier Oil Refining Corporation merged with Ashland (see Exhibit 2).

The fact that Ashland Oil was an "independent" in an industry dominated by very large firms makes the success more interesting. Furthermore, a large portion of the profits throughout its history resulted from refining operations—a branch of the industry in which it is generally considered difficult to maintain stable earnings over a long period of time. Naturally, faced with these facts, one would wonder just how this company happened to make this achievement.

Was this success merely the result of luck? Was the company riding a growth industry in which any firm would be able to make such a showing? Undoubtedly, as we will see later in the historical facts, chance cannot be ruled out altogether. Nevertheless, a comparison of Ashland Oil's profitability with that of other companies in the same industry, as charted in Exhibit 3, points to the conclusion that factors other than chance were significant determinants of this success. It will be noted in studying Exhibit 4 that Ashland's ratio of net income after taxes to capital and surplus was consistently higher than the ratio for eighteen of the major companies and also higher than the ratio for nineteen independents except for the war years. This exception is interesting, since the least dramatic success occurred during periods of governmental controls, when many of the determinants were out of the power of the company management.

Was this record by the company the result of the "intuition" and "innate ability" of Blazer? The story which develops in the following chapters indicates that certain qualities possessed by Blazer were necessary for the strategic actions. Yet the important question is whether prospective managers might be able to develop a "way of thinking" based upon concepts held by Blazer which likely would enable them to improve their own administrative ability. The concluding chapters will summarize these concepts and the important determinants of the achievements of the company and Blazer. This discussion should aid the development of better executives.

The reasonable conclusions derived from the historical facts in the following chapters are that the company was consistently

Name and Location of Corporation	Major Activity	Date of Purchase	Approximate Cost
Big Dipper Oil Company (Warren County, Kentucky)	Production	1919	$ 445,000
Quaker Oil Company (Lee County, Kentucky)	Production	1919	240,000
Great Eastern Refining Company (Ashland, Kentucky)	Refining	1924	250,000
Union Gas and Oil Company (Martha, Kentucky)	Production	1925	5,000,000
Highland Oil Company (Hillsboro, Ohio)	Marketing	1929	25,000
Home Oil Company (Maysville, Kentucky)	Marketing	1929	85,000
Southern Ohio Oil Service Company (Pomeroy, Ohio)	Marketing	1929	15,000
Dawson-Pepper Oil Company (Crooksville, Ohio)	Marketing	1929	27,000
Keaton Oil & Gas Company (Johnson County, Kentucky)	Production	1929	250,000
Economy Bulk Sales Company (Columbus, Ohio)	Marketing	1930	40,000
Tri-State Refining Company (Kenova, West Virginia)	Refining	1930	300,000
Cumberland Pipe Line Company (Home Office: New York City)	Transportation	1931	420,000
Rix & Corbin, Inc. (Lexington, Kentucky)	Marketing	1936	29,000
Giles Oil Company (Cynthiana, Kentucky)	Marketing	1936	90,000
Mt. Sterling-Aetna Oil Company (Mt. Sterling, Kentucky)	Marketing	1938	23,000
Allied Oil Company (Cleveland, Ohio)	Marketing	1948	11,500,000
Aetna Oil Company, Inc. (Louisville, Kentucky)	Marketing	1950	4,500,000
Freedom-Valvoline Oil Company (Freedom, Pennsylvania)	Marketing	1950	4,000,000
Frontier Oil Refining Corporation (Buffalo, New York)	Marketing	1950	2,600,000
Kellogg Petroleum Products, Inc. (Buffalo, New York)	Marketing	1954	1,350,000
R. J. Brown Company (St. Louis, Missouri)	Marketing	1956	1,000,000

* Includes only the more important acquisitions of Ashland Oil & Refining Company and its predecessors, Swiss Oil Corporation and Ashland Refining Company. When exchanges of stock were involved in an acquisition, the cost was estimated by using the current market value of Ashland stock received.

4

successful and that this success was the result of certain managerial concepts and operational policies directed by Paul Blazer. These concepts and policies provide interesting and unique information for the development of improved administrative thought and action. It will be shown that there are many ways in which Ashland's operations and managerial techniques differed from those of the larger companies.

PERSONAL BACKGROUND OF BLAZER

The chief determinant of these ideas and policies was the mind of Blazer, the chief executive. His life prior to joining the company, therefore, will help in the understanding of the nature of the man whose judgments directed the fortunes of the company.

Paul G. Blazer was born in New Boston, Illinois, on September 19, 1890, and spent his early life in nearby Aledo, where his father was a newspaper editor. While attending William and Vashti College at Aledo in 1908-1909, he promoted and operated a magazine subscription agency which provided him with an independent income. In 1910 he became manager of student subscriptions of Curtis Publishing Company in Philadelphia, Pennsylvania. In his boyhood, Blazer had obtained extensive advertising experience on his father's newspaper, which proved valuable in his future jobs. Although he was progressing successfully in his first executive position and although he had been offered a job at $100 per week by Crowell Publishing Company, he decided in 1914 to enter the University of Chicago, where he majored in social sciences and attended law school. He liked mathematics and physics but did not receive formal training in chemistry or engineering. Because war was imminent in early 1917, he dropped out of college just prior to becoming eligible for an undergraduate degree, enlisted in an army hospital unit, and married Miss Georgia Monroe in April, 1917. Dis-

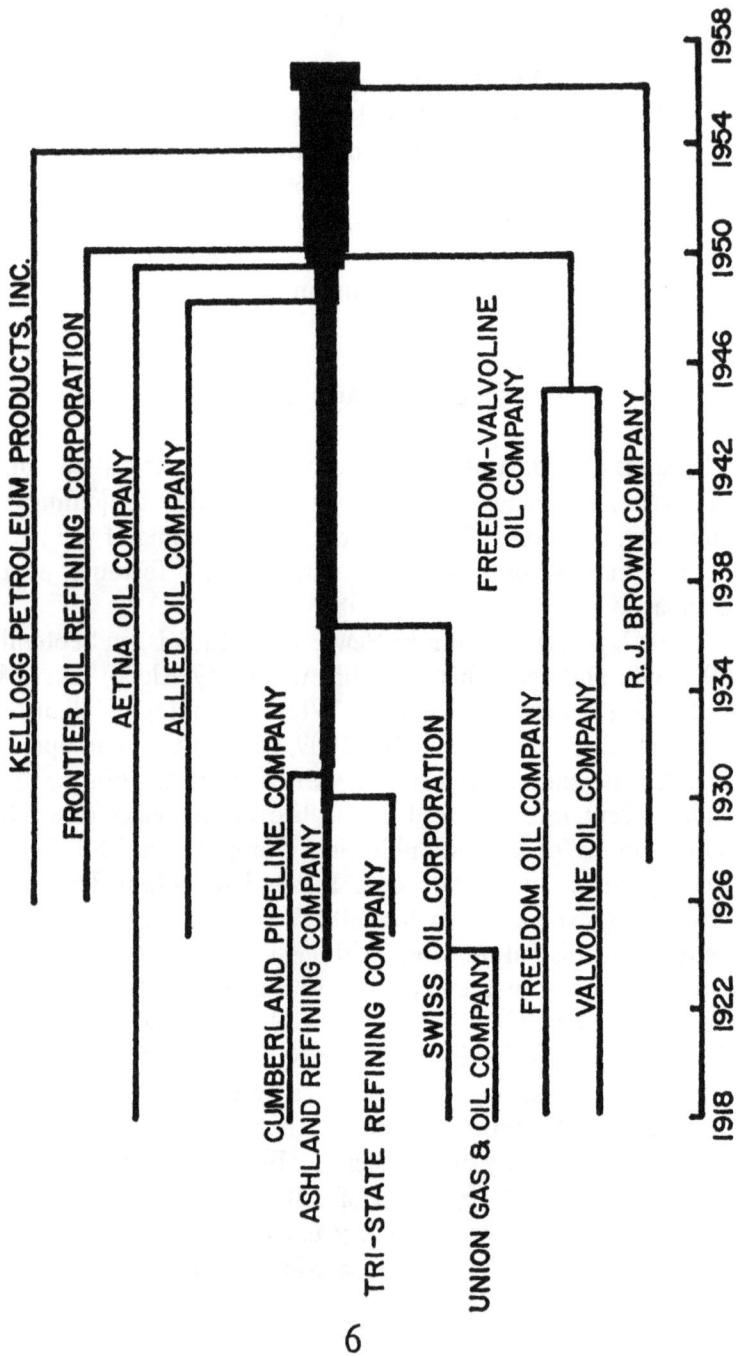

EXHIBIT 2—CHRONOLOGY OF ACQUISITIONS OF MAJOR ANTECEDENT COMPANIES, ASHLAND OIL & REFINING COMPANY

6

charged honorably from the military service as a result of a physical disability, he became advertising service manager of Chittendon Press, a Chicago printing company. In the role of advertising service manager for that company, he made his first contacts with the oil industry through the printing of an industry directory.

In 1918 Blazer joined a former schoolmate, J. E. Shatford, who had worked for the latter's father in a refinery at Blue Island, Illinois, and who had acquired a refinery at Joliet. Blazer soon became sales manager and vice president of the younger Shatford's firm, Great Northern Refining Company, which sold the output of both the Joliet and Blue Island refineries.

The need for additional crude oil to supply the Joliet refinery resulted in Blazer's first trip to Kentucky to investigate the Cow Creek (Estill County) area of eastern Kentucky and the Warren County area of western Kentucky. His purchasing activities resulted in the formation of the Great Northern Pipe Line Company. The quest was so successful that Blazer purchased more than enough crude oil and sold the surplus to other refining companies. Contacts made during these years of crude oil sales proved most valuable in future operations. Many acquaintances in the oil industry, referred to in his letters, date from these sales. Blazer's ability to forecast crude prices enabled him to anticipate changes in the crude market and to advise acquaintances wisely on the proper times to buy or sell crude. Personal relationships with oilmen, important in later negotiations, were strengthened in this manner.

In 1919 the Great Southern Refining Company was organized by the same financial interests. While employed by Great Southern in Lexington, Kentucky, Blazer obtained additional sales management experience, and during extended absences of the company's president, he acquired his first experience in refining management. He tried new methods in the simple refining process of that time and soon became known as a refining specialist, although he had had no engineering educa-

7

EXHIBIT 3

NET INCOME AFTER TAXES, ASHLAND OIL & REFINING COMPANY
AND MAJOR OIL COMPANIES, 1924-1956

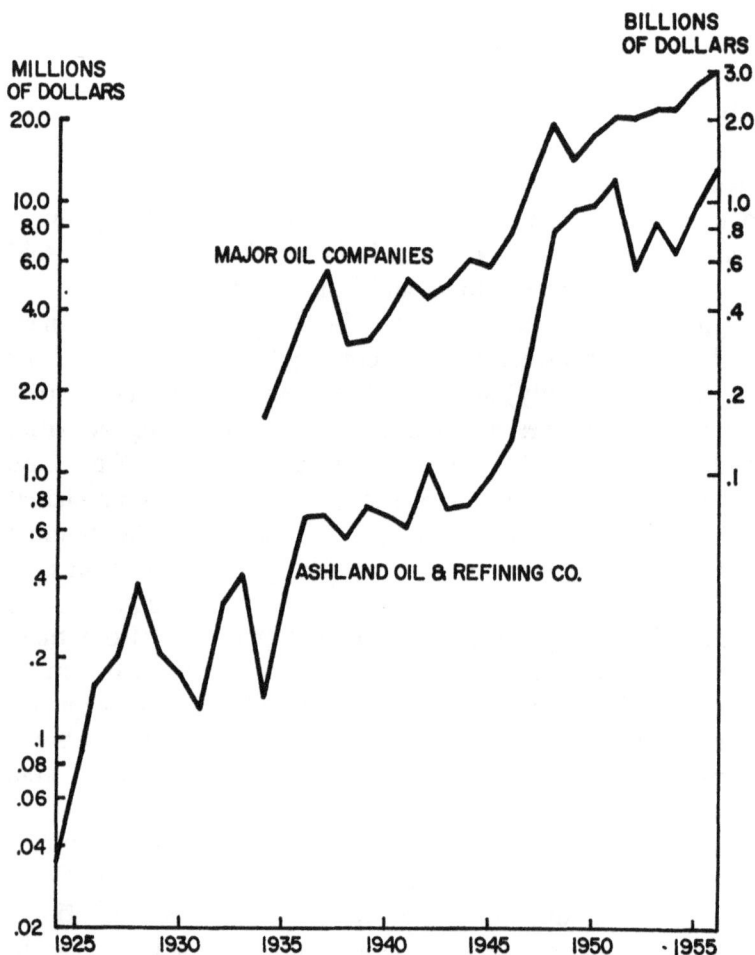

Source: Company records and *Financial Analysis of the Petroleum Industry* (New York, Chase National Bank, 1950, 1953, 1956).

tion. His interest in adaptations of refining techniques continued to be an important characteristic of his executive career. Blazer resigned his position with Great Southern to seek an association in refining in which he would own a greater financial interest. At the time of resignation, he was negotiating with Petroleum Exploration, Inc., Sistersville, West Virginia, which was considering integration into refining. Because unanimous approval by the board was required, a single dissent blocked the refining venture by Petroleum Exploration at Latonia, Kentucky.

On January 1, 1924, Blazer was hired by Swiss Oil Corporation to organize Ashland Refining Company, of which he became one of the incorporators. As partial incentive, Blazer was given an option to purchase 10 percent of the capital stock in the company.

NATURE OF THE PETROLEUM INDUSTRY

The general characteristics of the industry are especially important as determinants of the actions of the Ashland management. The fact that the industry has experienced continued growth makes it possible to assume that expansion is normal and even necessary. It enables some companies to continue to operate with techniques and policies which probably would cause difficulties in a more mature industry. This growth encourages managers who possess imagination, vision, and little respect for generally recognized methods of management to experiment with new techniques.

The dynamic nature of the industry demands that decisions be made quickly and that they be changed suddenly as a result of unexpected developments. Discovery of a new oilfield, a new process of refining, a new law, or a change in the pattern of transportation and marketing have often quickly altered the future outlook of a company.

The smaller the company in the industry, the greater the need for its management to adjust to external conditions. The

EXHIBIT 4
RATIO OF NET INCOME AFTER TAXES TO STOCKHOLDERS' EQUITY,
EIGHTEEN MAJOR COMPANIES, NINETEEN INDEPENDENTS, AND
ASHLAND OIL & REFINING COMPANY, 1929-1950

PERCENT

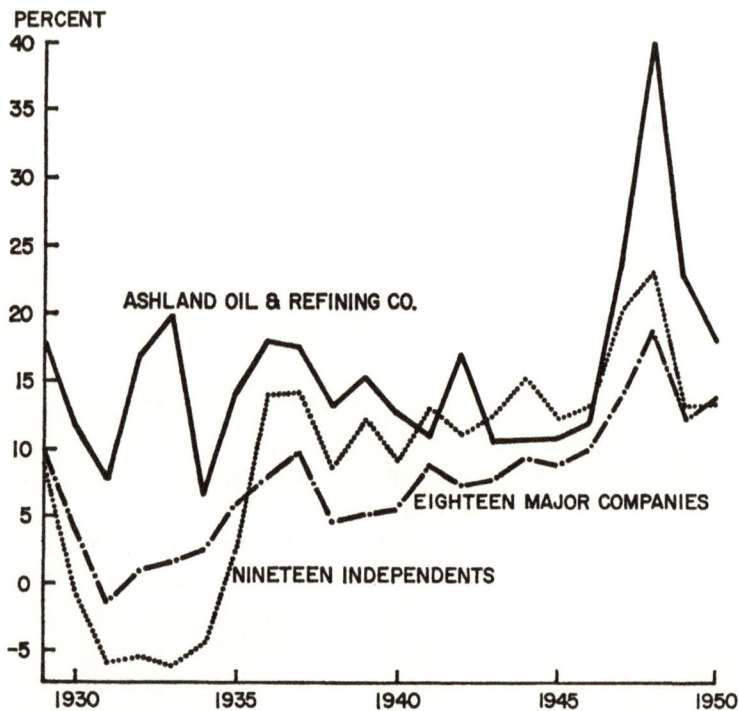

Source: Company records and Cassady, *Price Making and Price Behavior in the Petroleum Industry*, 328.

10

management of the smaller company usually cannot affect significantly the industrial pattern; but as a company grows and becomes a significant factor in the industry, it becomes better able to plan its own policies for the longer run with more assurance that those policies will affect the pattern of the industry. The complexity of the oil industry has caused many management problems. Some observers feel that the four major branches are separate industries in themselves. Most, however, consider production of crude oil, transportation of crude and refined products, refining, and marketing as major segments of a single industry. An integrated company has operations in all of these four branches. In many companies four major executives are chiefs of these segments; often operations in these segments are through separately incorporated companies. The scope of top management is complex and broad; a specialist in any one segment may have extreme difficulty in securing a broad outlook into all aspects. Furthermore, related activities are becoming more closely associated with petroleum: petrochemicals offer broad opportunities for expansion, and rubber and other industries have become allied through discoveries of new products and processes.

Several recent studies have shown the importance of vertical integration in the oil industry. McLean and Haigh explained: "Large, integrated units have gradually emerged as the predominant form of business organization in the oil industry because the economic climate in the industry has been generally favorable to the integration process for a long period of time."[1] The growth of integrated companies had its origin in the character of profit opportunities in the four branches, the characteristics of the physical facilities necessary in the process, the nature of the managerial job in conducting the activities, and the

[1] John McLean and Robert W. Haigh, *The Growth of Integrated Oil Companies* (Boston, Harvard University Graduate School of Business Administration, 1954), 663.

absence of serious obstacles to vertical integration. McLean and Haigh found that from 1920 to 1950 the proportionate share of the industry's capacity held by small firms declined from 28 percent to 15 percent. The means by which Ashland Oil countered this trend and later became large therefore offers an interesting area of study. The fact that all large oil companies and many small ones are integrated is important in considering the process by which Blazer sought the proper pattern of integration for his company. McLean and Haigh came to the conclusion that "there have always been many different integration arrangements in the oil industry which would permit profitable operations and effective competition. . . . It appears therefore that the formation of integration patterns is fundamentally a process of progressive adaptation by which *individual companies* work out the particular structures best suited to their *own particular* external and internal situations."[2]

The oil industry is composed of 8,000 to 10,000 producing firms, more than 17,000 wholesale distributors, almost 200,000 independently operated retail service stations, but fewer than 200 refining companies. Generally the structure of the industry is described as composed of "major" and "independent" oil companies. The term "major" usually applies to the twenty largest companies, which do approximately 75 percent of the total business; all the rest, including Ashland Oil, are "independents." Historically, the term "independent" referred to those companies not connected with the old Standard Oil Company which was dissolved in 1911, but the Temporary National Economic Committee in the late 1930's began to call the largest twenty "major." The terms thus are loosely used, and some so-called "major" companies, for example, Phillips Petroleum and Sun Oil, might argue that they should be classified as "independents." In any event, absolute size of the company appears to have much greater importance in organizational and managerial problems than does relative size.

[2] McLean and Haigh, 674.

12

The structure and competitive behavior of the industry has been under examination continually throughout its history. These investigations and the resulting legislation and court decisions have defined the legal setting that has determined the manner in which the Ashland management has reacted to a given set of circumstances. The federal government has given special attention to the encouragement and protection of the independent. Its efforts have often been discouraged by the actions of other independents. It is clear that the "independent producer"—the small, nonintegrated, crude oil production company—has interests that conflict with the "independent refining company"; often both have different viewpoints from that of the "independent jobber." On many occasions the divergent views have resulted in one group of independents testifying in behalf of the actions of the larger companies against other groups of independents. The larger firms have found that it is more effective to allow the smaller ones with common interests to lead. Some small firms have often found it good business to support the larger companies.

In this setting the Ashland management has consistently resisted the promotion of restrictive legislation to control the actions of the largest companies. It has felt that any attempt to freeze the relationships by a pattern of regulatory procedures would be to its own disadvantage; such attempts would eliminate the opportunities for shifting into new profitable areas. Although the management appeared to appreciate the beneficial effects of protective legislation on the opportunities of a small firm, it often supported the viewpoint of the larger companies. The apparent paradox was resolved in the pragmatic appraisal of each separate case. Generally, Blazer felt that independents could benefit more from the danger of governmental action than from the action itself.

13

In order to explain the evolution of Blazer's managerial philosophy, it is necessary to consider the characteristics of petroleum.[3] The fluid nature of the commodity facilitates materials handling. The ability to transport crude petroleum and refined products by pipelines, tankers, and barges sets the stage for integration into transportation through company-owned facilities. This liquid characteristic makes a continuous process in production desirable and demands that close coordination be maintained among the various stages of production.

The ease of movement, the high value of the commodity, the tendency to evaporate, and the high cost of storage space above ground necessitates continued close attention to the control of inventories. The organizational structure of an oil company must provide for means by which this flow of crude oil from the well to the consumer can be regulated and adjusted to the circumstances of the moment. In order to obtain security of both the source and the outlet of this flow, many firms have directed their efforts toward better balance in their integration.

The fact that oil is concealed in the ground and will flow to another's property, if not developed by an owner, has important effects on the problems of crude oil supply. Uncertainty in the exploration phase prescribes that a firm which desires to operate on a systematic basis with a secure supply of crude oil from its own wells must work with a long-range plan and use the law of large numbers to substitute a calculable factor for an indeterminate risk.

The chemical construction of hydrocarbons creates a potentially infinite number of different products which can be produced from crude oil. The various processes of treating the hydrogen and carbon elements lay the foundation of a dynamic industry. Such chemical characteristics make refining theo-

[3] P. H. Frankel, *Essentials of Petroleum: A Key to Oil Economics* (London, Chapman & Hall, 1946).

14

retically a problem of alternative products (variable quantities of different products can be obtained from a given barrel of crude oil) rather than of joint products (fixed proportions of each fraction must be produced from a barrel of crude oil). Practically, the potential changes in proportions is limited by technological problems. Greater flexibility can be purchased if the manager desires to place additional equipment in his refinery. The greater this flexibility, the greater is the choice that the manager has in the quantity of each fraction that he can produce. His problem then becomes more complex in that he must not only estimate the demand for a single product but also the relative fluctuations in the demands for a number of alternative products. He must make both short-run and long-run decisions relating to his correct actions for a particular situation.

The different proportions of the various fractions are partly determined by the different grades of crude oil. Petroleum is far from standardized: some types are suitable for a high percentage of gasoline; some yield less volume but higher quality of gasoline; some contain foreign substances—sulphur, for example—and require additional processing to eliminate the undesirable elements. The base of the crude oil may vary from predominantly asphalt to primarily paraffin. Other characteristics of crude oil make the raw material of the industry very heterogeneous and less adaptable to a single standardized process.

AREA OF OPERATIONS OF ASHLAND OIL

Most of Ashland Oil's operations have taken place in the area bounded by the Mississippi River on the west, the Allegheny Mountains on the east, the Great Lakes on the north, and the Cumberland River on the south. During the greater part of its history the company has concentrated its operations within the three states of Ohio, Kentucky, and West Virginia. In this ter-

15

Exhibit 5—Area of Operations,

ASHLAND OIL & REFINING COMPANY

TORONTO

ROCHESTER

BUFFALO

DETROIT

ERIE

Lake Erie

CLEVELAND

AKRON

NILES

Pennsylvania

CANTON

FREEDOM

PITTSBURGH

WELLSVILLE

Ohio

FOLLANSBEE

COLUMBUS

TON

MARIETTA

CINCINNATI

West Virginia

ON

ASHLAND

KENOVA

● Principal Offices

Refineries

Product Terminals

Lake Tankers

River Towboats and Barges

——— Ashland Products Liner

- - - Common Carrier Products Lines

★ Sales and Operating Offices

ritory a few independent refiners and their distributors compete aggressively with a large number of the major oil companies.

This marketing area has certain unique characteristics. It is close to an economic ridge of transportation costs: the largest refineries in the area receive their crude oil supplies principally from the Southwest; and refined products come from the West by pipeline, from the Gulf Coast by river barges, and through products pipelines from refineries and terminals on the eastern seaboard. It is the competitive marketing territory which marks the highest points of transportation costs from the largest oilfields and refining areas. It is here that competition from the East meets competition from the South and from the West. A filling station on one side of the street may receive its gasoline from a large eastern refinery on the Delaware River, and the station across the street may be supplied from a refinery in Oklahoma or Texas.

Although several oilfields are relatively close to this market, their production is insufficient to supply the demand for oil products: the specialized crude oil of Pennsylvania has declined in volume since its discovery in 1859, thus eliminating it as an important source other than for the manufacture of lubricating oils; eastern Kentucky has had production for a number of years but has remained an unimportant source relative to other areas. The largest source nearby is the Illinois basin, located in southern Illinois, western Kentucky, and southern Indiana, the importance of which increased rapidly in 1937 as new fields were discovered. The development of flush production in 1938 in the Illinois basin was of fundamental importance to the growth of Ashland Oil in the late 1930's. The extremely rapid development of these fields and the rapid production decline of each well caused a peak production to be reached in June, 1940, and a decline in total production thereafter in spite of continued drilling activity.

A most important characteristic of the location of Ashland Oil is the availability of water transportation. As Exhibit 5 illustrates,

18

the Ohio River flows through the center of operations, making possible low-cost barge transportation. In addition, the Great Lakes provide a waterway along one side of the territory. During the early development of crude oil transportation, pipelines were constructed from the Southwest into and through the area. The operation of many of these lines as common carriers permitted the growth of small refineries.

The market for petroleum products in this area is concentrated and capable of absorbing the many products from a barrel of crude oil. The density of population and automobile registrations causes the area to have a large consumption of gasoline. The northern temperatures make domestic fuel oil an important product. The large number of manufacturing plants and railroad shops offer a market for industrial lubricants. Industrial firms within the area also receive naphthas of varied types as raw material for their products. Asphalt for highways and residual fuel oil for steel mills are important uses for the heavier fractions.

Often in each market area of the oil industry there has been a reference marketer. Usually this marketer has been referred to as a price leader, although studies have indicated that while he may lead prices up, he characteristically follows them down. The comment is made often in the petroleum industry that a certain area is "Indiana territory" or "Jersey territory." Ashland's area of operations falls within parts of a number of different "territories": the reference marketer in Ohio has been the Standard Oil Company (Ohio); in Kentucky and Tennessee, Standard Oil Company of Kentucky; in West Virginia, Standard Oil Company (New Jersey); in Pennsylvania, the Atlantic Refining Company; in New York, Socony-Vacuum Oil Company; in Indiana and Illinois, Standard Oil Company (Indiana).

Ashland Oil has always operated in areas in which these large companies had a part in establishing reference points around which other firms planned their actions. This fact has put a premium on concepts and policies that would permit Ashland Oil to adjust to these multiple reference marketers with a

19

minimum of confusion and a maximum of attention to each local situation. If the company had operated in an area in which there was only one major marketer, the problem of maintaining attention would have been simplified and subject to greater standardization. The need to adjust to one set of conditions established by one company in one territory and to a second set of conditions established by a second company in a second territory has determined in part the managerial behavior of Ashland Oil.

This characteristic of adjusting to multiple reference marketers made it possible for Ashland Oil to remain insignificant in the area of each marketer and, therefore, to avoid some of the threats of retaliation that might have resulted from attracting too much attention by a large increase in volume in any single area. The strategy that evolved from this characteristic was one of shifting expansion plans from one territory to another in order to decrease the chance of significant results being felt by any single marketer. In other words, the attempt was to keep from getting too great a share in any single territory.

In 1956 Blazer looked back upon his experiences in the management of Ashland Oil and explained to a meeting of the Newcomen Society the way he looked at the challenges which faced a company which had grown as rapidly as his:

These fields of human relations are deserving of, and are receiving, greatly increased attention from all-corporate managements. As a company grows, the complexity and the importance of these relationships increase. In an expanding industry, however, continued corporate growth and skill in the solution of related problems appear to be necessary for competitive survival. The degree of success of management in meeting these interesting challenges appears to me to be of even greater significance than the growth of the business.

As Ashland Oil has continued to grow, so we have struggled to retain the informal, friendly organization of a small company. We have endeavored to inject into our relations with the public, our employees, customers, and stockholders, what I like to think of as a small-town, large-family attitude! That becomes increasingly difficult

20

now that we are owned by 30,000 stockholders, and operate in hundreds of communities through a number of thousands employees, with sales of $250,000,000 per year.

The chapters which follow describe the nature of this growth. Chapter II is, in a sense, a preliminary history of the company from which the refining firm developed. It provides most of the discussion of exploration, development, and production of crude oil. Chapters III and IV relate the history of the refining operations and activities which are the heart of the book: Chapter III covers the operation of the small Ashland Refining Company; Chapter IV describes the beginnings of Ashland Oil & Refining Company as it developed into a medium-size organization.

The period of extremely rapid growth through mergers is the subject of Chapter V. This chapter outlines the background of each of the larger companies which became a part of Ashland Oil and analyzes how these companies fit together. Chapters VI and VII respectively bring together the transportation and marketing activities. They also finish the historical aspects by describing the activities of the large oil company in the 1950's.

The primary biographical section, begun in this chapter, is completed with the analysis in Chapter VIII. Paul Blazer's managerial philosophy is summarized and evaluated in Chapters IX and X.

2

WILDCATTER IN KENTUCKY

The Operations of Swiss Oil Corporation

OIL PRODUCERS in the United States looked forward to a very prosperous year in 1918. The wartime demand for petroleum products had jumped the price of crude oil from less than $1.00 a barrel in 1915 to approximately $2.50. Too, the number of motor-driven vehicles on the highways was expected to exceed six million. Gasoline had already replaced kerosene as the chief product of crude oil, and the thermal cracking process discovered in 1912 by Dr. William M. Burton was beginning to show its importance for automobile fuel.

The promise of easy money encouraged wildcatting-speculative drilling in unproven oilfields—during 1917 and 1918, and men with big dreams and little capital eagerly flocked to wherever a new strike was reported. To Kentucky they came after Charles Dulin found oil on Tick Fork of Cow Creek, near Irvine, in 1916. Oil had been struck in the state as early as 1819, although there had been no considerable commercial production until the 1880's, and there had been mild booms in 1900 and again in 1912. In the few years after Dulin's find, however, oil exploration pushed Kentucky's production to a peak of more than nine million barrels in 1919. If the finds were never as spectacular as those of Oklahoma, which was also experiencing an oil boom, the risks were fewer; the Kentucky oil sands were much nearer the surface than the Oklahoma pools, and consequently less capital was required to drill a well.

J. FRED MILES AND THE FORMATION OF SWISS OIL

The requirement of less capital attracted J. Fred Miles, a wildcatter who had been operating in Oklahoma, to Kentucky in 1917. He traveled throughout eastern Kentucky purchasing oil leases in untested areas. On May 15 he contracted to buy fortynine producing wells in Estill County from United Oil Company. To search the extremely complicated titles to the leases for these wells, he employed a Louisville lawyer, E. L. McDonald. When McDonald found the titles to be defective, the Cleveland financial interests who had considered backing Miles decided against proceeding further. Miles' original effort to promote an oil-producing company died for lack of outside support.[1]

On a second attempt a year later, Miles made contacts with men who possessed sufficient venture capital and succeeded in incorporating Swiss Oil Corporation in Kentucky on June 21, 1918. This was the second corporation using the name Swiss, the first being organized as Swiss Drilling Company in Oklahoma by Miles in 1910. Leases in seven eastern Kentucky counties were transferred to the new Kentucky corporation. J. I. Lamprecht, president of National Refining Company, Cleveland, was elected president, and Miles became treasurer and general manager. E. L. McDonald, who has remained an officer in Swiss and Ashland Oil for more than forty years, was named secretary soon after formation. Since Lamprecht, who was interested in the Kentucky corporation primarily as a supplier of crude oil to his refining company, did not attend a Swiss board meeting, Miles was the principal driving force.

Miles expanded Swiss' operations into the Big Sinking field

---

[1] Further information concerning this initial effort by Miles can be found in *Miles v. United Oil Co.*, 192 Ky. 542 (1921); 204 Ky. 346 (1924).

in Lee County during the first months of 1919. A sawmill was constructed near Fixer on Bobby's Ridge to supply lumber for the camp—fifteen three-room houses, a schoolhouse, a church, a commissary, a "flicker," a large headquarters house, and a baseball diamond. The nearest railroad was at Torrent, twenty-two miles to the northeast, to which transportation was provided by several span of horses purchased from a brewing company. There were no roads in the area, and often block and tackle were used to carry heavy items from one ridge to another.

Labor during this period was scarce. Most of those hired were unskilled workers who came from surrounding eastern Kentucky counties. Total employment in 1919 increased from ten to approximately twenty-five men.

Membership on the Swiss board of directors changed rapidly. During the summer of 1919, Miles interested a number of Chicago investors in the company's operations. After James L. Martin had visited Fixer to report on the possibilities of the company, H. J. Halle, Martin J. Insull, Arthur Reynolds and George M. Reynolds of Continental and Commercial National Bank, J. L. Washburn, and later, Ogden Armour bought Swiss stock. These men were to play a major financial role in the company during later years. Martin, who became a director on July 10, 1919, remained on the board until his death in May, 1954. He was elected chairman of the board of Ashland Refining Company in 1935.

On March 1, 1919, the first dividend was paid on the Swiss stock. During the following months of 1919 and 1920, earnings of the company supported a liberal dividend. As a result, the company was able to obtain additional capital, with which Miles expanded his drilling operations into Warren County, Kentucky, two hundred miles to the southwest of the Lee County field. A well known as "Big Moulder," drilled earlier in the year near the Barren County line, was such a fabulous producer that Miles purchased it along with a pipeline and other equipment on July 10, 1919.

24

The purchase of Big Moulder probably had a lasting effect on the future management of Swiss. Because Miles spent a relatively large part of the new funds obtained from the Chicago financial interests on a single well which soon after purchase produced only a fraction of its expected output, he was later unable to maintain his prestige with the stockholders. In addition, his construction of an expensive gasoline plant in Lee County hurt his reputation. Events eight years later showed the effects of this lack of confidence.

At the end of 1919 the Big Sinking field was booming; price of Somerset crude oil, which it produced, was $3.25 a barrel; assets of Swiss Oil Corporation were valued at $2,700,000; eleven leases and 131 wells were producing earnings, slightly more than enough to cover dividend payments of $215,275. Expansion continued through the purchase of Quaker Oil Company, which owned nearby leases in Lee County. Charles E. Evans, one of the larger stockholders of Quaker Oil, served continuously on the board of Swiss and Ashland Oil.

In 1920 Swiss experienced its last boom. Little of the profit was retained in operations; the general manager's salary was increased and a short-term loan was obtained. Lamprecht died and John Gund, owner of a Lexington, Kentucky, brewery, was made president of the company on October 16, 1920. The rapid change of events is tersely described in the words of the annual report for 1921:

During the latter part of 1920, the opinion was freely expressed that consumption was overtaking the supply, and that there was necessity for increased production. As a result, the development of oil resources and search for new pools were stimulated to an unprecedented extent. In January, 1921, Somerset Light Oil was selling at $4.50 a barrel, and a prosperous year for the oil producer was in prospect. An unexpected decline due to over supply and various contributing causes drove the price down in the course of a few months from $4.50 to $1.00 per barrel.

This sudden change of the price together with the very

25

EXHIBIT 6

SWISS OIL CORPORATION

B A L A N C E S H E E T

DECEMBER 31, 1921

A S S E T S

CURRENT:

Cash on Hand and in Bank		$ 53,265.20
Accounts Receivable and Notes Receivable:		
Stockholders and Employees	$17,152.07	
Swiss Pipe Line Company	8,000.00	
Trade and Other Accounts	13,943.46	39,095.53
Inventory of Oil in Storage Tanks		5,570.34
Total Current Assets		$ 97,931.07

INVESTMENTS:

Swiss Pipe Line Company Stock		83,100.00

PROPERTY:

Producing Properties	$3,297,514.46	
Drilling Tools and Machinery	21,091.09	
Furniture and Fixtures	3,442.46	
Gasoline Plant	239,948.55	
Pipe Line	23,379.88	
	$3,585,376.44	
Less: Reserves for Depletion and Depreciation	647,319.72	
	$2,938,056.72	
Nonproducing Leases	265,431.28	
		3,203,488.00
Total Assets		$3,384,519.07

L I A B I L I T I E S

CURRENT:

Notes Payable	$ 346,000.00	
Accounts Payable	74,235.95	
Reserve for Federal Taxes	67,355.78	
Total Current Liabilities		$ 487,591.73

CAPITAL STOCK ISSUED AND OUTSTANDING

		2,800,000.00

SURPLUS:

Appreciation of Lease in 1919	$ 131,300.00	
Paid in Surplus	180,160.00	
Earnings from Period of June 21, 1918		
to December 31, 1921	343,562.52	
	$ 655,022.52	
Less: Dividends Paid during years		
1919 and 1920	558,095.18	
		96,927.34
		$3,384,519.07

serious current financial condition at the end of 1921 (for example, current assets totaled $97,931.07, and current liabilities, $487,591.73; see Exhibit 6) would have terminated the company if it had not been for the financial connections in Chicago. Not until April, 1922, however, were the short-term liabilities refunded into first mortgage 7-percent sinking fund gold bonds with a maturity date of May 15, 1927. In the handling of this issue, Robert D. Gordon, Chicago investment specialist, first became interested in the company.

The problems of the oil industry were serious during these early years of the 1920's. In the annual report for 1921, Miles continued his explanation of the factors which affected the company's condition:

The great depression . . . was largely the result of the combined operation of the following, viz.

1. The reaction after inflated war conditions from which no branch of industry has been exempt.

2. The excessive drilling campaign in 1920 induced by high price of oil and prediction freely made of oil shortage resulted in discovery and development of large new pools.

3. Decline in industrial demand for fuel and lubricating oils and heavy ends of refined products.

4. Stagnation in export demand for all petroleum products.

5. Unprecedented heavy imports of oil from Mexico.

Total production of oil in the Big Sinking district declined because few new locations for drilling were available to compensate for the normal production decline of the existing wells. The annual report for 1922 summarized the conditions:

Our principal producing properties are in Lee and Magoffin Counties, Kentucky. Those in Lee County have been almost completely developed by drilling to the extent that is considered profitable, with crude oil at present prices, and our best properties in Magoffin County have only a limited number of locations to be drilled.

On June 5, 1922, the second president of Swiss died. He was succeeded by Senator Thomas A. Combs of Lexington, Kentucky, who was then president of Fayette Home Telephone Company and Combs Lumber Company. The fortunes of the oil company were at their lowest ebb. In 1923 losses ran to $259,250.49, which represented not quite 10 percent of the total asset value of the company. The trend was downward; it was obvious that something had to be done quickly. Many of the stockholders advocated liquidation of the company, but Miles and Combs searched for other possibilities.

REVITALIZATION OF SWISS OIL

At the meeting of the Swiss board of directors on January 28, 1924, two propositions were advanced by the management which were of such proportions that it might be maintained that the company began in 1924, not in 1918. Swiss Oil Corporation was, in fact, all but dead. Certainly the life of the company could not have been prolonged without some enormous change. That change occurred, curiously, at a time when the company had very limited resources.

The first proposal was the purchase of Union Gas and Oil Company of Martha, Kentucky, for a price not to exceed $4,500,000. This Lawrence County company had been incorporated under the laws of Indiana on March 21, 1917, by A. B. Ayres of Fortville, Indiana, Frank M. Millikan of Indianapolis, and A. C. Albin of Newman, Illinois. On July 8, 1920, an office was opened at Paintsville.

During the early twenties the three owners expanded drilling operations as fast as possible, but they found that their funds were limited, although they were relatively wealthy. Previous to this venture they had been primarily gasmen finding oil on the properties where they were looking for gas. The stockholders of Union found themselves confronted with numerous legal difficulties primarily resulting from failure to drill on leases within the proper time. Other oilmen were buying "top

28

leases" from the owners of the land and were successfully taking away valuable oil rights. So in 1923, when a Philadelphia firm offered $4,250,000 for Union, they decided to sell their properties.

Why should the owners be willing to sell? The answer to this question was that in spite of their realization that the properties were worth much more than the final selling price, they considered the following factors: the company had been organized originally for the production of gas, and the owners knew little about oil; the immediate drilling on a number of leases was a prerequisite for holding the leases; numerous suits being brought against the company were causing legal obstacles; and there was clearly disagreement among Ayres, Albin, and Millikan which made a situation where one of the owners found himself in the minority opposing the other two. Ayres, who owned 40 percent, had been willing to sell on the last account alone.

After several months of negotiations with Swiss representatives, on July 26, 1924, the stockholders of Union gave an option for the purchase of all the stock of the company at a price of $5,000,000. An initial payment of $1,500,000 in cash was to be made; the vendors were to retain control of the operations of the company until another $1,000,000 could be accumulated out of the net earnings; and the remaining $2,500,000 was to be paid at the rate of $200,000 per quarter. The option was renewed several times, but the Union stockholders had indicated that extension would not be made past January 12, 1925.

The properties under question were appraised by J. I. Lamprecht, II, president of National Refining Company, at approximately $12,500,000. Even with this appraisal it appeared to be impossible to interest any of the investment houses who investigated the financial requirements, for Swiss Oil Corporation, with small net assets and a poor earnings record, was attempting to purchase a company with assets many times its own.

As it might be expected, Combs, Miles, and McDonald had a

difficult time interesting any financial groups. Starting in 1923, they had negotiated with Shaffer Oil and Gas Company, which later decided against the venture. Then an attempt was made to merge with Union, but two of its three owners, Millikan and Albin, declined the offer. Combs and Miles next traveled to New York and Cleveland in quest of financial backing. C. N. Manning and J. D. Van Hooser of Security Trust Company, Lexington, accompanied the Swiss representatives on some of these trips. A. D. Hambleton & Company of Baltimore and Blair & Company of New York showed some interest, but neither would handle the deal. Cyrus Eaton of Otis and Company, Cleveland, seemed ready to make an offer at one time, but he failed to follow up with a definite commitment.

After this "nibble" from Otis, a local group in Lexington made a definite offer; however, the directors of Swiss felt that it was not attractive and refused it at a board meeting on January 10, 1925. At this meeting Combs stated that unless the offer could be exercised, he would recommend that the company liquidate its affairs, for it was losing money and it would only be a question of time until its entire capital would be lost. It was then that James L. Martin, of Pynchon & Company, made a move to investigate whether his partners would help finance the transaction. Since Pynchon had eleven partners scattered in Chicago and New York, Martin was forced to make rapid contacts. On the last day of the option, confirmation was obtained which made possible the purchase of the Union properties. To make the notes marketable, it was necessary to issue $2,000,000 of 7-percent mortgage notes and $2,000,000 of common stock of Swiss Oil Corporation in order to obtain $1,750,000 in cash. Later in 1927 a lawsuit was brought by one of the stockholders of Swiss claiming that the loan was illegal under Kentucky law, alleging that the $2,000,000 in common stock was given without due consideration. The case was decided in favor of the defendants on November 25, 1927, and affirmed in an appeal on May 10, 1929.

WILDCATTER IN KENTUCKY

The record of the decisions of the judges offers important information concerning the condition of Swiss Oil Corporation on January 12, 1925:

At the time of the transaction the defendant corporation faced liquidation. Without the acquisition of the stock or property of the Union Gas and Oil Company it could not continue in business. . . . Every day which it continued to operate under existing conditions, it was getting farther and farther behind. This is not disputed.

. . . . .

Prior thereto, the stock of the defendant corporation was selling at from ten to twenty per cent of par value, whereas after its acquisition in 1926, its stock, according to Miles' testimony, was worth par. . . . Evidently the defendant corporation is in a flourishing condition notwithstanding the depression in the oil market.

. . . . .

It clearly appears to have been, if not the universal, at least the general opinion of the officers and directors of the Swiss Company that it had no recourse but liquidation unless it acquired the Union Gas and Oil Company properties. . . . The Swiss Company could not have borrowed $2,000,000 in the usual way. Its own property was already mortgaged. It appears that its first mortgage bonds, issued several years before, had been underwritten at 90 per cent of their face.

. . . . .

True, several of the directors of that Company bought from Pynchon & Company substantial amounts of the bonds at but little above their cost to Pynchon & Company. But as we understand the record, such sales were intended not as a means of securing the contract from the Swiss Company, but of insurance to that extent of Pynchon's ability to sell the securities at some profit.[2]

At the time of the Union purchase the Chicago interests gained majority control of Swiss. R. D. Gordon became a director and has remained active in the affairs of the company for more than thirty-five years. The large powers once held by Miles, the original promoter and general manager of the company, now were limited by the board of directors. For the next

[2] *William H. Lamprecht II* v. *Swiss Oil Corporation et al.*, 32 F. (2d) 646 (6th Cir. 1929).

31

fourteen months Miles battled to regain his previous control. Then on March 21, 1927, a new superintendent was employed to fill the vacancy created when Miles left to start a refining company in Louisville.

Until he was forced out, Miles had used his close relationships with National Refining Company to strengthen his position. As a boy in Oklahoma, he had met the elder Lamprecht, president of National Refining Company. Lamprecht was not only the first president of Swiss Oil Corporation, but National was a most important customer of Swiss. During the financial crisis of 1921-1922, advances from National Refining Company helped Swiss carry its large current liabilities. The brief of the defendants in *Lamprecht II* v. *Swiss Oil Corporation* indicates the internal relations within Swiss Oil Corporation during the two-year period, 1925-1927:

It came to the attention of the directors during the summer of 1926 that the stockholders were being circularized for the purpose of getting options on their stock, and at the director's meeting on July 24, 1926, the matter was brought up and Mr. Miles was asked to state whether he had any connection with it, and whether Mr. Berry, who was sending out letters and option agreements to stockholders, was acting under the instructions of Mr. Miles. Mr. Miles said he had no connection with the matter; and that he would send a letter to all the stockholders stating this fact. Accordingly such a letter was prepared by the President [Combs] of the Corporation, but he [Miles] refused to sign it, and so far as the evidence discloses he never did send such a letter.

. . . . . .

It is respectfully submitted that the evidence given by Mr. Miles at the trial fully discloses that he was engaged in an attempt to place the control of the corporation in the National Refining Company, the Corporation's principal customer at that time and as long as Mr. Miles was in the saddle and exercised complete control over the Company's business and affairs, the National Refining Company felt no need to invest in the Company's stock, but shortly after his activities were curtailed and his powers subordinated to those of the president, the National Refining Company considered the best interests would be served by buying into the Company.

32

Until the annual report dated March 3, 1925, Miles had signed all company reports. After that date and until 1935, Combs, the president, signed them. The date of the curtailment of powers of Miles, the former general manager, was significant in the affairs of Swiss Oil.

The organization of the Ashland Refining Company was the result of action of the Swiss board of directors on January 28, 1924; Paul G. Blazer was named general manager of the new firm. At the time, this event appeared to be much less important than the Union purchase, which was approved at the same meeting of the Swiss board, for it involved an expenditure of only $212,000, in contrast to the $5,000,000 price for the Union company. This small transaction was reported in the minutes of the board of directors of Swiss Oil Corporation as "negotiations . . . toward the purchase of the plant of the Great Eastern Refining Corporation, located near Catlettsburg, Kentucky, from which it appeared that said plant could be purchased at the price of $212,000 to be paid $50,000 in cash and the balance in five installments."

Thus the management of Swiss Oil Corporation decided to integrate forward into refining. It was this action in 1924 that gave birth to a subsidiary which within twelve years became the chief operation.

PERIOD OF DECLINING PRODUCTION

After the purchase of the Union properties and until 1942, the company did little to increase its oil reserves. It drilled a few wells, but not enough to offset the decline in production. The relatively large capital expenditures for production in 1930, evident in Exhibit 7, were for secondary recovery by repressuring with gas the properties obtained from Union.

From 1925 to 1935 the producing properties were managed by Combs from Swiss' home office in Lexington, while the

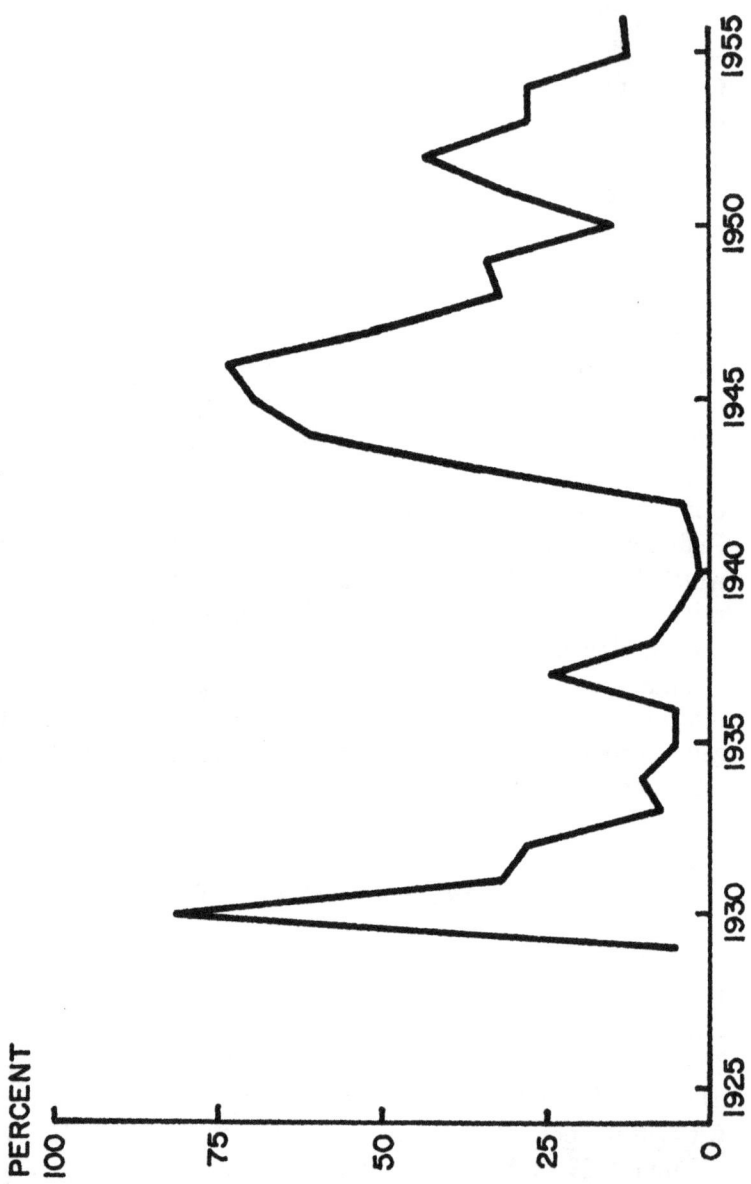

EXHIBIT 7—PERCENTAGE OF TOTAL CAPITAL EXPENDITURES PLACED INTO PRODUCTION OF CRUDE OIL, ASHLAND OIL & REFINING COMPANY, 1924-1956

34

growing refining operations were supervised by Blazer in Ashland. Blazer kept the Swiss management informed by lengthy letters and reports, but he experienced little interference in his management of the Ashland operation. Blazer had very little to do with the producing branch, which then was completely separate from refining, both geographically and organizationally. Except as a member of the Swiss board after 1929, Blazer was not concerned with production of crude oil at the time when he was having great success in refining, transportation, and marketing. Later, in 1935, when the offices of Swiss Oil Corporation were moved to Ashland, the producing branch was a stepchild. It was the only part that had not been built by Blazer.

Exhibit 8 shows the actual net production of crude oil from 1924 to 1956. From a high of 2,675 barrels per day in 1926, net production dropped gradually throughout the 1930's. The only interruption was the slight upward effect of the repressuring operations of 1930 and a small increase in 1938. Exhibit 9 shows that in 1925 Swiss' crude oil production was 155 percent of Ashland's crude runs to stills, whereas it was only 24 percent by 1938. This drop was the result of the slow but steady increase in refinery runs.

After 1927, crude oil prices were not high enough to encourage major emphasis on production. The price of Somerset grade of crude oil produced in eastern Kentucky dropped from $2.30 a barrel in 1927 to $0.90 in the early 1930's. The tremendous production from the newly discovered east Texas fields made the operation of the small wells in eastern Kentucky uneconomical.

By July, 1930, Ashland was purchasing 75 percent of the crude oil carried by Cumberland Pipe Line Company, which collected most of the crude in eastern Kentucky. Ashland thus was able to obtain more than adequate supplies from independent producers. Blazer saw that Swiss was losing money in production; therefore he reasoned that deemphasis on production was logical. In fact, at times he found that excess supply of crude oil was a problem. He was aware, however, that he must offer

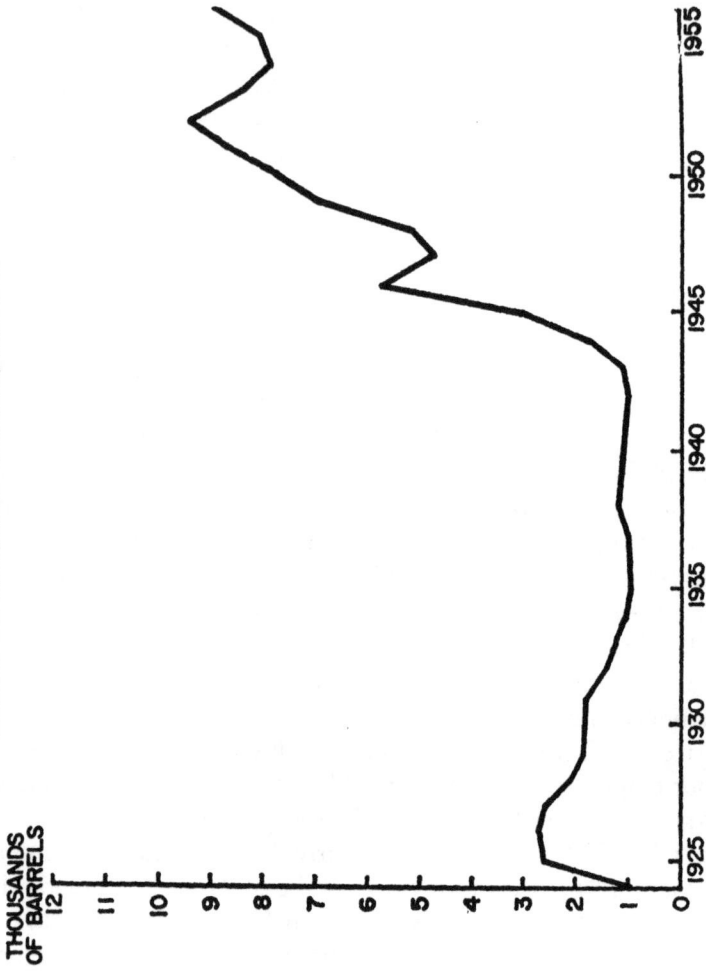

EXHIBIT 8—AVERAGE DAILY NET PRODUCTION OF CRUDE OIL, ASHLAND OIL & REFINING COMPANY, 1924-1956*

Source: Company records.

* Includes Swiss Oil Corporation; excludes interest in foreign production.

a market for independent producers in order to keep them as a potential future source of supply. He expressed his views to the Swiss president in 1930:

As a result of this emergency, I would recommend that the Swiss Oil Corporation, as the largest producer of Eastern Kentucky crude oil, curtail production in so far as that can be done without benefiting your neighbors or injuring your properties.

. . . . .

As I see it, the present situation endangers the Ashland Refining Company only in so far as it threatens the prosperity and continued operations of Eastern Kentucky producers whom we must look to for further supplies of oil. I am not worried on our own account, but since we are the only important present outlet for Eastern Kentucky crude oil, heavy responsibility rests upon us.

This sense of responsibility to the independent producer was maintained through Ashland's history and was one of the reasons the company could continue to obtain sufficient crude oil from independent producers in spite of the low percentage of company-owned production.

Although Seep Purchasing Agency, representing South Penn Oil Company, originally a part of the old Standard Oil Company, had for many years posted the official market price for Eastern Kentucky crude oil, Ashland had become the principal purchaser and refiner of Kentucky oil and began posting its own market price, which usually was higher than the price paid by Seep. In February, 1931, Seep Purchasing Agency discontinued the purchasing and posting of a market for Kentucky crude oil, and thereafter the official market price was posted by Ashland. The Texas Company, the only other important purchaser and refiner of Kentucky crude oil, accepted Ashland's postings. The reason for this change in the pricing of Kentucky crude oil was Ashland's ability to accept all the oil that was offered to it and to pay more than anyone else.

This leadership in crude purchasing fell to Ashland early in 1931 within three months after the Ashland management had

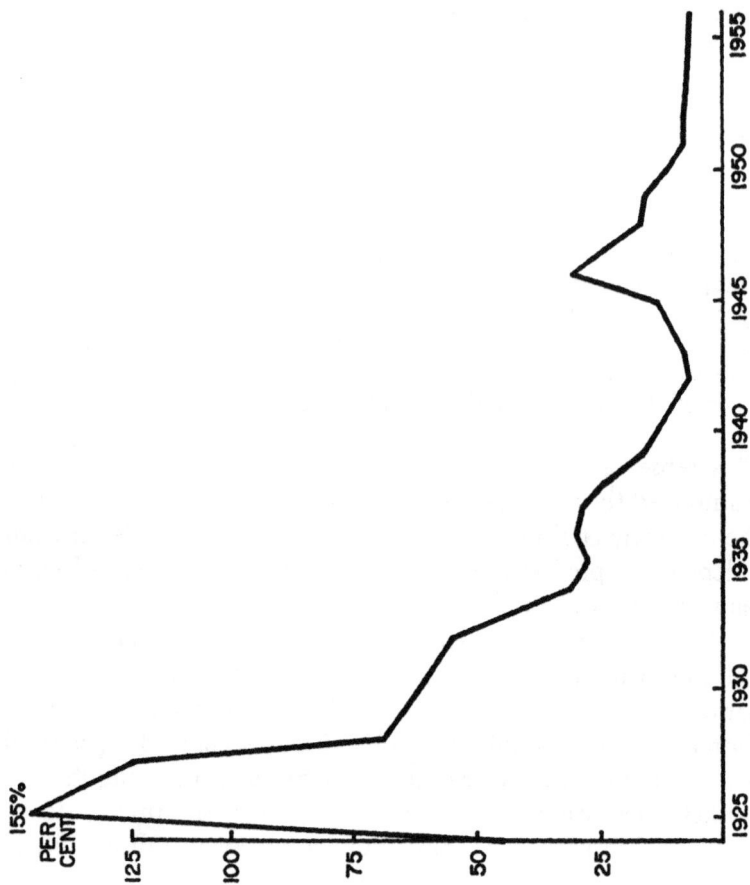

EXHIBIT 9—RATIO OF NET PRODUCTION OF CRUDE OIL TO REFINERY RUNS TO STILLS, ASHLAND OIL & REFINING COMPANY, 1924-1956*

Source: Company records.

*Prior to 1936, the net production of Swiss Oil Corporation was considered with the crude runs to stills of Ashland Refining Company.

38

attempted to sell the producing properties to several of the large oil companies. In the last three months of 1930 the financial pressure on the Chicago interests in Swiss was so great that they asked Blazer to sell the producing properties. For two months Blazer talked with representatives of Cities Service Company, Sinclair Refining Company, and others. Since all had more crude oil than they needed, no satisfactory sales could be arranged. If these negotiations had succeeded, the history of the company would have been significantly altered. Ashland would have had no ownership in producing fields and so would not have been in a position to make its strategic purchase of a pipeline one year later.

The price of Swiss stock was so low in the early 1930's that it would have been easy for an investor to gain control of the company. This was the period, however, of severe stress in financial markets. During this unsettled situation, William C. Freeman of Chicago first learned that there was such a firm as Swiss Oil Corporation, and in 1935 he began his long membership on the board of directors. One day a banker brought this financial specialist a basket of certificates representing part of the collapsed Samuel Insull empire and requested Freeman to see if he could realize any value from them. In that pile of paper were shares in Swiss Oil Corporation.

The greatly devalued stock on Freeman's desk reflected the continuing financial difficulties of Swiss. In his letter to Combs on January 9, 1934, Blazer referred to the 1931 situation:

The present favorable financial condition of the Swiss Oil Corporation contrasts most favorably with the situation which I recall prevailed three years ago this week. There were outstanding at that time Swiss debenture notes totalling $778,000 which were maturing on the 15th of the month. Due to prevailing disturbed financial conditions, the sale of the new issue of Swiss bonds had not met with the degree of success anticipated. Our companies, as a group, were already borrowing $845,000 from banks, which was the extreme limit of their credit. Default of the maturing notes, which would have been disastrous for the sale of the new bonds, was threatened and doubtless would have occurred except that you pledged your own credit and that of the Combs Lumber Company to raise an additional $100,000 for the Swiss Oil Corporation; also, on the day before the notes were due, another large stockholder and director, following your example, pledged his personal credit to raise approximately the same amount to make up the remaining deficit, thereby enabling the company to weather the storm. Probably few of your stockholders knew of the danger that threatened at that time, and probably an even smaller number of your stockholders know that only a few years previously the directors of the company had debated as to whether it might be better to liquidate the corporation while there were still sufficient assets to pay off the creditors. Instead, the officers and directors took the only alternative—they plunged more heavily into debt and purchased new properties.

In January, 1933, the par value of the capital stock of Swiss was decreased from $5.00 to $1.00; the assets of the company were written down by $2,225,000 to reflect the lower market values; a capital surplus was created for the purpose of making dividends possible. In 1934 Swiss extinguished its long-term indebtedness and paid a liquidating dividend from capital surplus. In a letter to stockholders dated June 15, 1934, Combs, the president, communicated the glad news:

You will be interested to know that your company has paid and retired every dollar of its bonded indebtedness, 18 months before final maturity, bonds in the principal sum of $430,400 having been paid since January 1, 1934. Your company has not, at this time, any bank borrowings nor does it owe anything other than small current accounts.

The management of Swiss now began to consider a possible merger with its subsidiary, Ashland Refining Company, whose earnings by 1935 were approximately ten times those of the parent producing company.

On April 7, 1935, Thomas A. Combs died. In the consequent reorganization, James L. Martin became chairman of the board of Swiss and Blazer became president. For the first time Blazer held operating responsibility for the production of crude oil.

Combs had entered the management of Swiss when it appeared that nothing could be done for the company. He had successfully revitalized the producing company, supported and encouraged Blazer in his operations of Ashland Refining Company, and paid out the entire indebtedness of Swiss at the bottom of the Great Depression. His death removed the last real reason to maintain separate companies in two locations. On October 31, 1936, Swiss Oil Corporation and Ashland Refining Company were consolidated.

3

ECONOMIES OF OBSOLESCENCE

The Operations of Ashland Refining Company

THE ALMOST offhand manner in which the board of directors of Swiss Oil Corporation decided to integrate forward into refining belies the careful work in preparation for this action. It reflects, however, the directors' trust in Paul Blazer, who was named general manager of the newly organized Ashland Refining Company.

Blazer and Ashland more than fulfilled the board's expectation of substantial returns on its investment in the refinery. Although the initial operations of Ashland Oil & Refining Company were in Swiss' production of crude oil, the stable foundation for its future growth was laid in 1924 with the refining activities. Throughout the thirty-three years after that date, the Blazer management based all of its decisions to integrate backward into transportation and production of crude oil, and forward into transportation and marketing of the refined products, upon the assumption that the chief source of profits was refining.

CHARACTERISTICS OF REFINING

Refining is the manufacturing stage of the petroleum industry. Its nature is determined by the varying chemical qualities of crude oil and the processes required to segregate the fractions or components of the refined products. Furthermore, the demand for various fractions fluctuates greatly. Add to these factors dynamic and rapid technological change, and it is not surprising that successful refining operations require management to watch closely the numerous variables and to adjust operations to unique sets of conditions.

The processes which the refining manager must direct and control may be divided into three groups: distillation, alteration, and purification. Distillation is the basic process which involves heating the crude in a still and running it through a fractionating tower; as the gases cool, they condense on various trays within the tower, depending upon the boiling point of each fraction. The fractions are then piped off as different products—gasoline, kerosene, fuel oils, and so on. When Ashland came into being in 1924, refining involved little else than a simple distillation.

Techniques of alteration have been developed in the industry in the last forty years. Thermal cracking was the first to affect the operation of Ashland; later, catalytic cracking became a necessity as the required quantity and quality of gasoline increased. Both processes involve the breaking of heavier hydrocarbon molecules into lighter ones: heat is the agent in the case of thermal cracking; a foreign substance is used in catalytic cracking. Polymerization and alkylation are other advanced methods of alteration; in these processes the lighter gases are treated in order to make heavier fractions of high-quality gasoline. Alteration can also be achieved by use of additives.

Numerous types of purification are needed to eliminate undesirable elements which may be present in crude oil. Some crudes have a high sulphur and salt content which, if not removed, causes excessive corrosion. Lubricating oils require special processing and purification.

Continuous operation in all processes is made possible by the fluidity of the product. The need for large amounts of heat makes economically mandatory a work schedule of twenty-four

43

hours a day, seven days a week. Large amounts of capital equipment and relatively few laborers are required. Operating personnel must remain alert in order to control the processes; maintenance personnel must keep the equipment in operating condition and aid in the periodic cleaning process known as "turnaround."

Management of a refining company depends upon several economic characteristics of the operations. Bain summarizes these factors in the following manner: "The principal economic aspects of refinery operation include the magnitude of the cost of refining, the manner in which this cost varies with the scale and character of facilities, the degree to which the variety of physically different refined products obtained from crude are subject to conditions of joint supply and cost, and the manner in which costs are apportioned among refined products."[1]

The cost of the complete manufacturing operation in the oil industry is small relative to the cost of crude oil, cost of transportation, and cost of distribution. Bain has observed that "efficiency and cost control in refining are not necessarily prerequisite to survival in the petroleum industry." Of course, Bain's statement applies to the integrated company. A firm which attempts to specialize in refining, where the margin is relatively small and unstable, must be very efficient. The refining specialists must adopt measures, including managerial techniques, which will offset the disadvantages of nonintegration. A company which has operated profitably over a long period in the refining branch primarily must have developed operational and managerial techniques especially suited to its circumstances.

The short-run costs determine certain operating policies of a refining firm. Variable processing costs, including fuel, supplies, and direct labor, are small relative to total fixed costs. The refiner, given a plant of a certain size, usually desires to maintain capacity operations at all times, because the incremental proc-

[1] Joe S. Bain, *The Economics of the Pacific Coast Petroleum Industry* (Berkeley, University of California Press, 1944), pt. I, 84-85.

44

essing cost near capacity is insignificant as compared with the total unit cost of products refined (including the cost of the crude oil). The nonintegrated refiner treats his crude oil costs as variable; even a small cost disadvantage in the purchase of crude oil (for example, 10 cents per barrel) will be of major importance to his total variable costs. The integrated refiner looks upon a large part of his crude oil costs as fixed and does not consider them of importance in the short run. The policies, therefore, that control the operation of an oil company will depend, in part, upon its degree of integration.

A study of long-run costs of refining introduces the complicated problem of the optimum size of a refinery. Location of a refinery in a small city such as Ashland, Kentucky, may yield wage differentials in favor of the smaller refinery. The delivered cost of the crude oil increases when extra amounts must be secured from more distant fields. The market area surrounding the refinery might be able to support only a relatively small throughput.

Ashland's management has weighed the variables in its adjustment to long-run cost factors. In addition, Ashland Oil has specialized in capitalizing on opportunities to buy small refineries at bargain prices. Special operating practices appear to have been efficient in the small-scale plants.

During Ashland's thirty-three years of operations, the factors favoring larger scale have multiplied. The introduction of pipe stills and continuous cracking by 1927 caused the optimum scale to increase; the minimum size of crude distillation units necessary to supply feed stock for a catalytic cracking unit was an added factor forcing a larger plant by 1940; increased use of pipelines to deliver products from the refinery increased the market area to a size that would support a large-scale refinery after World War II; the development of petrochemicals calls for even larger plants.

Important aspects of the economics of refining are the concepts of common or alternative costs and joint costs. The distinction between the two is important in a study of the development of operating policies. As the economist defines the term, true joint costs exist when two or more final products emerge from a productive process in proportions which are absolutely invariable. Alternative or common costs exist when two or more final products emerge from a productive process in proportions which can be varied at the discretion of the manager.

In the early methods of refining, especially in straight distilling, the fractions flowed from the process in a fixed proportion; the only way in which the manager could vary the quantity of his refined products was by altering the definition of each fraction. New processes have continually been discovered by which the various fractions can be reprocessed and additional quantities of other fractions obtained. Theoretically, the refiner could continue to combine or break down the molecules in a given crude oil so that he could attain almost complete variability among the products produced. Economically, however, many of the presently known methods of chemically treating the fractions are impractical. If sufficiently complicated trays, "side-draw pipes," and special processing equipment are included in the design of a refinery, it is possible to obtain increased variability. A fundamental aspect of Ashland's operational policies was the attempt to make more products alternative rather than joint *in the short run.*

This technical refining potential will be referred to later as one aspect of technological flexibility. A small refinery without cracking facilities must produce products under almost true joint cost conditions and is therefore limited in the application of this concept. A large refinery with the necessary technical equipment finds that the managerial problems of coordination distract attention from producing small quantities of specialty products. The medium-size company, theoretically, can acquire the necessary equipment and at the same time can give executive attention to smaller quantities of specialty products.

But in 1924 Blazer's first concern was not with specialty

products. His immediate task was to derive profits from a refinery which had been operating at a yearly loss of almost $50,000.

The refinery purchased by Ashland Refining Company and financed by Swiss Oil Corporation in 1924 was at Leach, Kentucky, on the Chesapeake and Ohio Railroad approximately six miles up the Big Sandy River from its confluence with the Ohio River at Catlettsburg.

The location at Leach had been considered for a refinery by several groups prior to 1922. The president of Great Southern Refining Company had given thought to it in 1921. Possibly some of the first ideas about a refinery at that location were those of John Buckingham, president of an Ashland bank, and Paul Blazer, who was then vice president and sales manager of Great Southern.

The refinery had actually been constructed in 1922 for Great Eastern Refining Company. John C. Kelly and J. A. Dalton, who were primarily coal operators, were the chief owners of Great Eastern. In spite of the fact that the plant was technically efficient, Kelly and Dalton lost $96,325 in less than two years. As a result of these unprofitable refining operations and their additional financial problems in the coal business, Swiss was able to buy the plant for $212,000 in cash plus the market value of the inventories.

The purchase was the result of a combination of factors. First, the unprofitable operations and financial difficulties of Great Eastern set the stage for possible negotiations. Second, Swiss Oil Corporation was finding it difficult to secure markets for its crude oil. Combs and Miles had been looking for some operation which could help reverse the declining trend in their company's business. The refinery at Leach was located near a common-carrier pipeline which could be used for cheap trans-

Exhibit 10

NOVEMBER 1-1923 AVERAGE RETURNS PER BARREL
OF KENTUCKY CRUDE REFINED AT VARIOUS POINTS.
****

	CASH YIELDS		COST OF CRUDE & MFG.		NET PROFIT
(1)					
Louisville,	Gasoline 31x9.75	- 3.02	Market	1.46	
using Cumber-	Kerosene 8x6.75	- .54	Premium(average)	.13	
land Pipe Line	Gas Oil 10x4.75	- .47	Pipeage	.25	
& rail trans-	Fuel 47x3.75	- 1.76	Transportation	.41	**22¢ loss**
portation	Per gallon	5.79¢	Mfg. & Sales	.40	
	Per barrel	$2.43		$2.65	
(2)					
Cincinnati &	Gasoline 31x10	- 3.10	Market	1.46	
Latonia, using	Kerosene 8x6.75	- .54	Premium(average)	.13	
Cumberland Pipe	Gas Oil 10x5	- .50	Pipeage	.25	
Line & rail	Fuel 47x4	- 1.88	Transportation	.41	**13¢ loss**
transportation	Per gallon	6.02¢	Mfg. & Sales	.40	
	Per barrel	$2.52		$2.65	
(3)					
Lexington,	Gasoline 32x9.50	- 3.04	Market	1.46	
using Cumber-	Kerosene 8x6.125	- .49	Premium(average)	.13	
land Pipe Line	Gas Oil 10x4	- .40	Pipeage	.24	
& water trans-	Fuel 47x3.31	- 1.55	Transportation	.15	**8¢ loss**
portation.	Per gallon	5.48¢	Mfg. & Sales	.40	
	Per barrel	$2.30		$2.38	
(4)					
Pryse, using	Gasoline 32x9.25	- 2.96	Market	1.46	
Cumberland Pipe	Kerosene 8x6	- .48	Premium(average)	.13	
Line. (Typical	Gas Oil 10x4	- .40	Pipeage	.24	
of Great South-	Fuel 47x3.25	- 1.53	Transportation	.00	**2¢ profit**
ern situation)	Per gallon	5.37¢	Mfg. & Sales	.40	
	Per barrel	$2.25		$2.23	
(5)					
Catlettsburg,	Gasoline 32x9.75	- 3.12	Market	1.46	
using Cumber-	Kerosene 8x6.75	- .54	Premium(average)	.13	
land Pipe Line.	Gas Oil 10x4.75	- .47	Pipeage	.24	
	Fuel 47x3.875	- 1.82	Transportation	.00	**27¢ profit**
	Per gallon	5.95¢	Mfg. & Sales	.40	
	Per barrel	$2.50		$2.23	

portation from the Swiss and other Eastern Kentucky wells to the refinery. Third, Blazer, who had previously made numerous contacts in the oil business in Illinois and Kentucky and who had obtained experience in refining operations while working for Great Southern Refining Company, was available for its management.

When purchased, the refinery had a capacity of between 1,000 and 1,500 barrels of crude oil per day. Its greatest advantage lay in its favorable location. This factor was clear to the management at the time of the purchase and remained important to successful operations over the ensuing thirty-three year period. In November, 1923, prior to the purchase, the most important advantages were analyzed in a report to the Swiss board by Paul G. Blazer, summarized as follows:

1. Leach, Kentucky, is close to a marketing area which experiences the highest prices for finished products. The reason for these high prices is that oil flows east from the large fields of Oklahoma and Texas and meets the oil flowing west from Terminals on the East Coast at points along a line extending from Utica, New York, through Johnstown, Pennsylvania, to Charlottesville, Virginia. The areas around this line or "economic ridge" are those of highest transportation costs. If a firm could operate along this line at low costs, it is obvious that it would have a large competitive advantage.

2. Leach is close to the crude oil supply of Eastern Kentucky. Availability of crude oil would be a minor problem and the cost of transporting the crude oil from the well to the refinery would be low if the common carrier pipe line was used.

3. Industrial plants in the area will make it possible to dispose of the fuel oil produced in a skimming plant at advantageous prices.

4. The existence of river transportation from the refinery will make it possible to extend the marketing area along the river and still keep transportation costs low.

5. Sufficient skilled labor is available within the immediate market area.

With this lengthy report was a tabular comparison of the hypothetical average returns per barrel of crude oil refined at five possible locations: Louisville, Cincinnati, Lexington, Pryse, and Catlettsburg (Leach). Exhibit 10 is a reproduction of this table. The estimated 27 cents per barrel profit clearly would have been superior to that of the other locations and large enough to tempt a producing company to integrate forward. This estimate proved to be high for 1924 and 1925, in which the profit per barrel actually earned was 8 cents and 14 cents respectively. Adverse conditions resulted in an inventory loss during the first half of 1924, improvement took place in 1925, and by 1926 the actual profit per barrel of 25 cents was near the estimate of November, 1923.

Ashland Refining Company began business on February 11, 1924. Blazer as general manager was assisted by only one full-time operating executive, Charles A. Jouett, the first refinery superintendent. Total employees numbered twenty-five, not including five clerical and supervisory personnel. In a short time this small group trebled the throughput which Great Eastern had attained. Major improvements were postponed, but attention was given to small individual economies. The first operations used a batch method of distilling, whereas modern refineries had adopted a continuous process. Ashland's management attempted to guard against those practices which had caused the early problems of Swiss—primarily, overextension of financial resources. This policy was mandatory, since the parent could ill afford to invest additional funds in its refining venture at the time of the Union Gas negotiations.

ONE-MAN MANAGEMENT

In his first annual report to the Swiss board, Blazer delineated several operational policies that were to continue throughout his management. There was strict credit control, which minimized the requirements of working capital; for example, at the end of the first year, a total of only $53.00 in accounts receivable was older than one month. Ashland made sales in carloads lots to minimize the cost of small direct shipments; barges served buyers located adjacent to river delivery points. The company depended upon independent jobbers as its sales outlets. Attention to small details, especially those relating to costs, was characteristic of initial operations. The management cooperated closely with independent producers in order to establish greater security in crude oil supply. Close attention to the level of inventories provided further security of raw materials supply and decreased risks from price changes.

The major equipment of the original plant consisted of four batch shell stills. Although they were of a design that other refinery managers were changing, the company secured a gasoline yield of 30.48 percent from 450,000 barrels of crude oil in 1924. This compared with an industry average for that year of 31.2 percent.

Refinery prices were unstable. It was not uncommon to be operating at a sizable loss in one month and at a profit in the next. Since, even at this early date, gasoline contributed more than half of the realization per barrel of oil, it became urgent for Ashland to increase gasoline yield. The inflexibility of the original equipment made it difficult to adapt quickly to new conditions. Blazer discussed this situation in his report to the board of directors in 1926:

Due partly to general conditions but primarily to the fact that fuel oil represents a larger part of our finished product than of our competitors, we made our best showing during the winter months when fuel oil was in demand and lost money during the summer months of high gasoline consumption.

Ashland's equipment in 1925 yielded slightly more than 30 percent gasoline, while the industry averaged 32.4 percent. By 1926 the industry's average had increased to 34.9 percent. Consequently, Blazer submitted to the Swiss board in August, 1925, an illustration (Exhibit 11) of the effect of this yield upon

EXHIBIT 11

COMPARISON OF ACTUAL RESULTS OF REFINERY OPERATIONS USING ORIGINAL EQUIPMENT
WITH ESTIMATED RESULTS USING FRACTIONATING TOWERS, July, 1925.

Using Original Equipment

Revenue: Return per 100 gallons of crude

Gasoline & naphtha	$.1446	30.19 Gals.	$4.365
Kerosene	.0651	7.18 "	.467
Fuel & gas oil	.0431	58.99 "	2.542
Evaporation, etc.	.0000	3.64 "	
		100.00 "	$7.374

Realization per bar. crude oil $3.097

Cost:

Cost of crude including premium	2.638
Transportation of crude	.24
Cost of manufacturing, overhead, etc. (55,119 bar.)	.286
Cost per bar. crude oil	3.164

Loss per bar. crude oil processed using original equipment $ .067

Using Fractionating Towers

Revenue: Return per 100 gallons of crude

Gasoline & naphtha	$.143	34. Gals.	$4.862
Kerosene	.0651	4. "	.260
Fuel & gas oil	.0425	58.36 "	2.480
Evaporation, etc.	.0000	3.64 "	
		100.00 "	$7.602

Realization per bar. crude oil $3.193

Cost:

Cost of crude including premium	2.638
Transportation of crude	.24
Cost of manufacturing, overhead, etc. (55,119 barrels)	.266
Cost per bar. crude oil	3.144

Profit per bar. crude oil processed using fractionating towers $ .049

operating profits. These figures reveal that an increase of the gasoline yield to 34 percent would result in an increase of 9.6 cents per barrel in realization. Costs would be decreased slightly —2 cents per barrel—if fractionating towers were used. All other assumptions remained the same in both illustrations. Actually, however, the amount of loss due to evaporation was soon decreased from the 3.64 percent shown in the table to 1.86 percent in January, 1926. Both the estimates in the table were more conservative than the guarantees by the prospective builders of the equipment.

In step with rapid changes in the industry during 1925-1926, Blazer changed the distillation in two stills from a batch operation to a continuous process; the batch method was maintained in the two other stills. "Bubble" or fractioning towers were constructed for each of the stills in the continuous process. These improvements, completed early in 1926, cost approximately $100,000—almost 50 percent of the initial cost of the plant.

The very favorable earnings record of this early period resulted from a favorable location, an expansion of diversified lines of products, low manufacturing and overhead costs, and an increased gasoline yield. Blazer explained in a report to the board:

We are now having a more diversified line of gasoline and naphtha and can seek new markets thus minimizing our risk from price wars. Now that we can manufacture gasoline with proper specifications, we are giving our product trade names. For example, our Red Pepper aviation gasoline which sells at a premium of 5 cents per gallon is already meeting with great favor. Our new fractionating equipment enables us to more than double our production of high test gasoline for which we have a good market practically the year around.

A strong oil market helped earnings at times (for instance, 1926 and 1928); however, Blazer looked upon times of market weakness as potentially benefiting his company. In 1925 he commented: "Many refineries are already showing losses and some are planning to close down. That will doubtless help the situation." When marginal firms were forced out of the market, the more efficient firms would then have a better chance of operating at a stable profit.

Low manufacturing and overhead costs resulted from making small changes in operations as the situation demanded. Although Blazer was not an engineer and had little experience in refining prior to 1923, his emphasis upon small improvements rapidly gave him the reputation of being a refining specialist. His success in the development of more efficient heat exchangers was less dramatic than many later actions; yet it probably was one of the most fundamental reasons for large profits at times when other companies had difficulty in obtaining any return from refining. In a report to the board, Blazer emphasized the importance of this factor:

Our substantial increase in throughput can be attributed to our new system of towers and heat exchangers which outside refinery engineers report as being the most efficient they have seen. By utilizing what was previously waste heat, the temperature of our crude is raised to over 360 degrees and about half of the gasoline has been taken off before the crude goes to the first still.

Blazer gained valuable experience in making use of old equipment, and this enabled him to capitalize repeatedly upon the acquisition of equipment which others considered obsolete. Executive attention to small economies made it possible to increase the efficiency of operations primarily through attaining throughput always in excess of "capacity" and sometimes almost double previous efforts.

Blazer referred in a report to the board of directors to these small economies:

We are practicing economy in many small matters, which until recently have been considered of too little importance to bother with, but viewed over a period they amount to substantial sums. The expenditures for the equipment to obtain these economies are small in comparison with the ultimate savings.

Blazer recognized the tremendous technological advances being made in the industry. After Dr. Burton of Standard Oil Company (Indiana) developed his method of thermal cracking, other processes became available. By the early 1920's the Holmes-Manley process of the Texas Company, the tube-and-tank process of Standard Oil Company (New Jersey), the Cross process, the Fleming process, and the Jenkins process were in use by competitors. The Dubbs process, owned by Universal Oil Products Company, was available to independent refining companies.

By the end of 1926 Ashland's management had placed a Dubbs thermal cracking unit in operation at a cost of $150,000. Although numerous changes were made in the original equipment, this plant has remained important in the refining facilities of the company. The personal care and attention given the Dubbs unit during the twenty-seven years made possible its efficient use much longer than would normally be expected. The organization which made this possible was one key to the technological advance of the company.

Such management techniques of Ashland made it feasible to produce numerous specialty products which were too trivial for the large companies to consider and too technical for small refiners to handle. In 1926 Blazer reported on some of these products:

We manufacture six grades of aviation or high test gasoline and two grades of commercial gasoline. These various grades of gasoline are sold under many different trade names, central and southern Ohio affording us our greatest market. We make eight grades of naphtha which include paint blender's naphthas, rubber solvent, dry cleaners' naphtha, soap makers' naphtha, manufacturer's kerosene, special motor tractor fuel, domestic furnace distillates, black oil for mine use, road oil, ink oil, and seven grades of petroleum coke.

Until 1928 Ashland had little staff organization. Indeed, at one time Blazer found himself operating without any other line executive of any kind. In 1927 he remained at the refinery for three days continuously as the only major overseer when Miles, having purchased a shutdown refinery at Louisville, employed Ashland's refinery superintendent at a higher salary, who in turn took with him almost his entire staff of trained supervisors. For a short period, the company literally operated as a "one-man" enterprise.

Until operations became larger, Blazer negotiated all of the sales personally and made decisions concerning the types of specialty products that he could refine. Changes in prices of the numerous products caused him early to consider incremental costs of each product in relation to its potential price as the controlling factor.

The development of the diversified line of products, the shifting of market areas to yield a greater netback,[2] and the ability to make improvements to obsolete refining equipment that proved to be the best available at the time were factors which enabled the company to continue to make profits in spite of the instability of the petroleum industry brought on by new techniques which were rapidly making older refining methods obsolete and by discovery of new oilfields which caused wide fluctuations in the crude oil market.

The year 1926 was the last one of prosperity in the oil industry for a long time. Crude oil sold for $2.60 a barrel, the highest price since 1921 and higher than it would be until after World War II. In 1927 the drop in crude oil prices to $1.68 per barrel caused many producing properties to operate at a loss. At the same time, earnings of Ashland Refining Company continued to increase, from $186,230 in 1926 to $213,586 in 1927.

During this period of management by a single executive, Blazer was interested primarily in "staying competitive" in refining activities rather than in striving for greater integration. Swiss operated independently as a producing company; Ashland

---

[2] *Netback* refers to the net receipts of the company at the refinery gate after subtracting all of the marketing and transportation expenses.

Refining operated separately in the manufacturing branch. There was little coordination between the two except through correspondence between Combs and Blazer. Refining included the operations of a small river barge that delivered finished products to points within fifteen miles of the refinery.

By the end of 1928 the Ashland management had completed its major task of modernizing the Leach refinery. During this single year the throughput was increased from 47,000 to 120,000 barrels per month, raising the profit to $423,917, representing a return of 43.78 percent on owner's equity. Blazer reported to Swiss:

So far as our skimming plant is concerned it is not probable that we can obtain during 1929 further important improvements in either plant economy or quality of products. We have gone about as far as our type of distilling equipment will permit.

The company was to make no major improvement in refining for the next ten years. The management had obtained its Dubbs thermal cracking unit when it had appeared necessary. The next revolutionary developments in the industry did not occur until catalytic cracking appeared in the late 1930's. Blazer timed his major capital expenditures so that the new equipment would not rapidly become obsolete.

During these ten years, most of the top executives of the later periods were first added to the organization. Chapter 9 will analyze this evolution of the human element in the organization and show how the management constructed an organization that later was capable of rapid expansion. The "Ashland family" became a small, compactly organized group of personnel with little experience in the oil industry. Additional personnel hired at the depth of the depression were young and also inexperienced.

Beginning in 1927 and continuing for a number of years, the management received offers to purchase parts or all of the Swiss-Ashland companies. Because a minority of the stockholders were in favor of liquidating Swiss or selling the company as a whole, Blazer followed up the initial offers in a number of cases, although he personally did not want to accept them. He commented in a letter to a member of the board of directors on May 24, 1929, that "it is my opinion that a closely coordinated, economical, small operation such as we are gradually developing, can show larger earnings on the amount invested than if our refining and distributing operations were absorbed into a larger organization."

In October, 1928, an offer was made by the Texas Company to purchase Swiss' interests in the Big Sinking producing wells for $375,000. (These marginal properties eventually were placed under one ownership; in 1945 Ashland Oil purchased the holdings of the Texas Company.)

In August, 1928, after a new management had assumed direction of the Standard Oil Company (Ohio), talks were held preparatory to more formal negotiations relating to the sale of the Leach refinery to Standard. No actual offer was made because conditions in the Ohio company changed and made the sale impossible. Later, Standard acquired a refinery at Latonia, Kentucky, which could serve the southern portion of its marketing area. This refinery was the plant that in 1923 Blazer was preparing to manage as a subsidiary of Petroleum Exploration.

In June, 1930, Vacuum Oil Company showed interest in marketing facilities in Ohio which Ashland had acquired by that time. In August a feeler came from Shell-Union. The Standard Oil Company (Ohio) was named in further talks.

Starting in 1931 and from time to time for the next twenty-five years, there were rumors that Ashland was controlled or would become controlled by a major oil company. Standard of

58

Ohio was often specifically mentioned. Blazer reported that during the depression, customers and competitors would visit the refinery in order to determine its status, since it appeared unbelievable that a refining company could continue to make profits without becoming controlled by a large firm. He felt that larger companies had suggested to Ashland distributors that his firm was very insecure, causing them to worry about the possibility that they would find themselves later without a supplier. This uncertainty among the customers was no small problem for Ashland's management.

The belief that Ashland was controlled by another company probably resulted from reports about the various offers. Moreover, the large number of transactions that oil companies had with each other might have been interpreted as evidence that the companies were operated under a single control. During these years Ashland sold most of its gasoline in the wholesale market and often to large companies. There were many sale and exchange agreements with Standard of Ohio. Personal friendships and social relationships might have given further indications of an ownership connection. Blazer remained close friends with several officers of competing companies. Indeed, as a result of his personal friendships with executives in Ethyl Corporation and Standard of Ohio, relations between Ashland and some of the Standard companies were often as close as those effected between members of the Standard interests by their corporate heritage.

During the entire history of Blazer's operations in the oil industry, his personal acquaintances with leading men in the industry influenced his managerial techniques. Attendance at the various trade meetings enabled him to strengthen his prestige. Although no quantitative weight can be given to the importance of these personal relationships, it can be said that undoubtedly these contacts helped in times of crises. But Ashland at no time was in the control of any other oil company; Blazer was the ultimate source of all policies.

The first major purchase of another partially integrated company took place on April 9, 1930, when Ashland Refining Company acquired Tri-State Refining Company. Tri-State's refinery had been built in 1925 and improved in 1928 by the addition of a Jenkins thermal cracking unit. Tri-State had been the other principal competitive purchaser of Somerset crude from the Cumberland pipeline and had been the owner of service stations in Ashland's marketing area. The Tri-State purchase furnished additional cracking capacity and marked the first time that Ashland integrated forward into retailing of gasoline. The purchase was defensive as well as offensive, since Ashland's management feared the effects of possible purchase of Tri-State's facilities by a competitor.

In 1956 Blazer recalled this purchase in a speech:

One afternoon in 1930, I received a telephone call from the owner of the Tri-State Refining Company, which a few years earlier had built a new refinery of modern design across the Big Sandy River at Kenova, West Virginia, within sight of our refinery. He and I were close friends and keen competitors. He was inexperienced, however, in the business of oil refining and marketing. He told me that his plant recently had been losing almost $1,000 per day and that he would turn it over to us if we would assume and promptly pay off the accumulated indebtedness which he personally had guaranteed. He named an approximate figure of around $300,000 and we closed the deal over the telephone. I learned later that he had been trying to sell his refinery to one of the larger oil companies.

The annual report for 1930 by Swiss Oil Corporation to its stockholders summarized the facts of the purchase:

Through these two subsidiaries [Ashland Refining Company and Tri-State Refining Company] the Company now owns and operates two refineries within a very short pipeline distance from the Company's producing wells, one plant near Catlettsburg, Kentucky, refining about 4,000 barrels of crude oil per day, the other located at Kenova, West Virginia, across the Big Sandy River from Catlettsburg,

60

refining approximately 1,500 barrels of crude oil per day. Both plants are modern, having efficient "cracking" stills and low-cost operation.

The purchase of Tri-State Refining Company, made just prior to a sharp drop in sales of petroleum products, caused Ashland Refining Company, for the first time, to curtail its throughput because of insufficient sales. The decline became more severe, and by the first three months of 1931 the refining operations lost $33,000. Ashland's ability to meet a crisis through flexibility was especially evident at this time. Although the depression grew worse, the refining company was able to end the year with a net profit of $126,134 in spite of its first three months of losses. Like results were achieved on other occasions, although not to the same degree.

SUCCESSFUL OPERATIONS DURING THE DEPRESSION

The refining company continued to show substantial profits annually through the depression. Blazer stressed flexibility as a fundamental factor in his letters to Swiss' president:

We are able to shift some of our production, because of flexibility in our operations, from gasoline to kerosene and furnace distillates, which bring higher prices than gasoline; we are succeeding in shifting some of our gallonage from cut-price areas to other territory, which as yet, has not been seriously affected.

A second factor in Ashland's successful operations during the depression was its ability in obtaining tetraethyl lead for use in regular gasoline to maintain a competitive position in octane ratings. Blazer reported that in spite of the opposition of several large companies, Ethyl Corporation began to sell lead for regular gasoline: "You will be interested to know that with the exception of one shipment to the Standard Oil of New Jersey, we received the first shipment of the 'yellow' lead."

A third important aid came from the development of a method

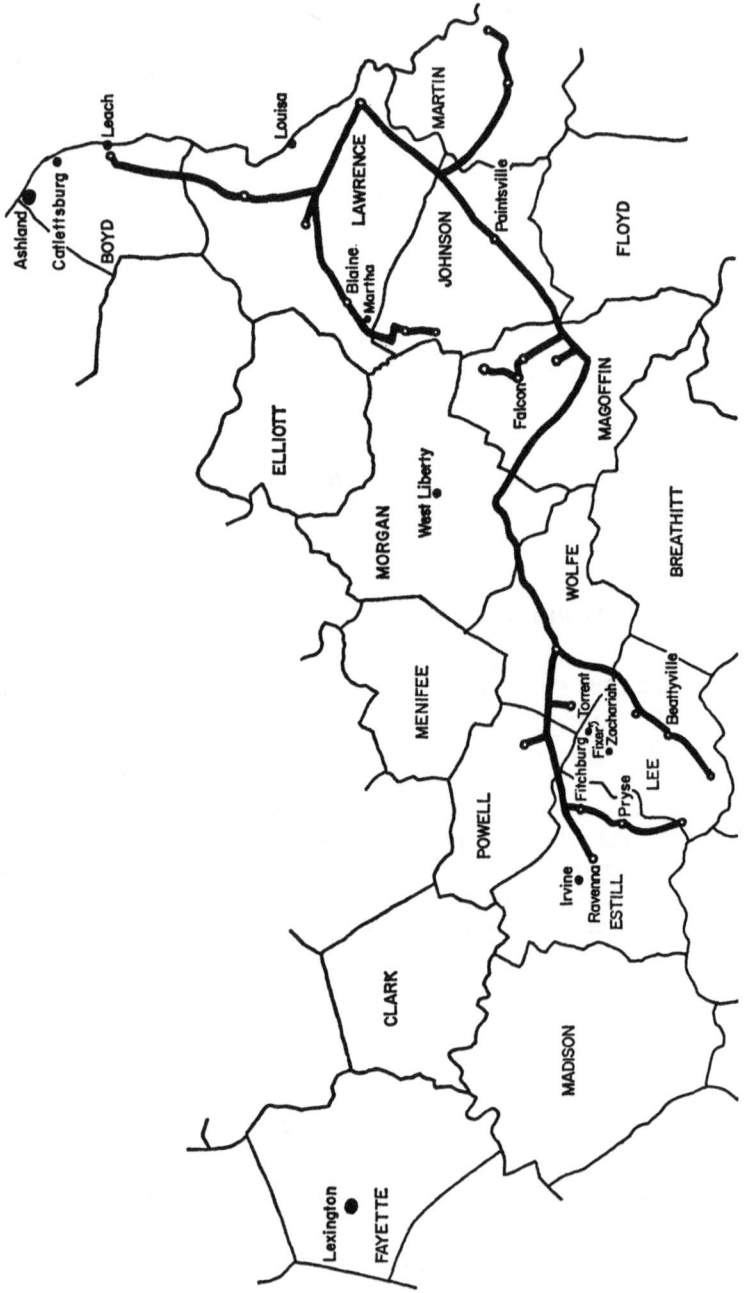

Exhibit 12—Geographical Location of the Cumberland Pipeline

62

of using the residue from the cracking process in the manufacture of a type of asphalt known as "cutback." After several years of trouble in meeting specifications, the company found the answers to its problems and asphalt became an important product. On May 19, 1934, Blazer explained in a letter that "our asphalt revenue last year accounted for a substantial part of our operating profit for 1933." The company organized a separate corporation, the Kentucky Stone and Tar Company, to promote its road-building activities.

A fourth favorable factor was the extent of integration which existed during these years. Blazer observed in a monthly letter:

Viewing our group of companies as a whole we are approximately 30 per cent integrated. We have slightly over 30 per cent of our crude supply and distribute through our own marketing facilities about 30 per cent of our gasoline output. . . . At times like the present, profits from the operation of the pipe line, our distributing operations and also from other "sidelines" cushion our refinery losses.

PURCHASE OF A CRUDE OIL PIPELINE

In 1931 eastern Kentucky crude oil flowed primarily to three refineries—Ashland's Leach refinery, Tri-State's Kenova, West Virginia, refinery, and Texas' Pryse refinery. The first two depended on Cumberland Pipe Line Company for a main artery from its crude oil supply (see Exhibit 12). This line had been started in 1902 by the Standard Oil interests. During the period of flush production, the line had operated at a profit, but with the decline of the oilfields, the firm, which was being managed from New York, found revenue diminishing. It appeared to Ashland's management that Cumberland operated with excess personnel and costly facilities. Furthermore, Blazer believed that dependence on Cumberland's tariff actions might have a serious effect on Ashland's profits.

Soon after the purchase of the Tri-State refinery in April,

1930, Blazer discussed with Cumberland's management the possible purchase of the pipeline. By this time Ashland was receiving 75 percent of the oil delivered by the line. Before approaching the Cumberland interests, Blazer had talked with representatives of National Supply Company concerning the probable cost of construction of a new line. With this possibility as a weapon, Blazer negotiated for the Cumberland line. He had been given the power to go much higher than the price at which he actually was able to buy the facilities.

The story of the negotiation with the Cumberland management was summarized by Blazer in 1956:

The Cumberland Pipe Line Company owned and operated the Kentucky pipelines of the Southern group of pipeline companies, which had a network of thousands of miles of gathering and trunk lines extending from the oil fields of Kentucky, West Virginia, and western Pennsylvania, to the large refineries located along the Delaware River. They were headed by Forrest M. Towl, located at 26 Broadway, in New York City, who as a young engineering graduate from Cornell University, had gone with Mr. Rockefeller in 1885, and had become chief engineer of the original Standard Oil Company. I thoroughly enjoyed my negotiations with Mr. Towl in connection with the purchase of the pipeline properties. He was truly a gentleman of the old school.

· · · · ·

The principal stockholder in Cumberland was the Rockefeller Foundation, which owned approximately 25 per cent. The balance of the stock was publicly owned, since it was traded on the New York Curb Exchange. Some of my friends and associates thought we were taking on more then we would be able to handle.

· · · · ·

Within a few weeks [after taking over the Cumberland line] we had put into effect many economies which turned the previous operating loss of Cumberland into a profit of more than $15,000 per month, thereby providing the installment payments as they came due. Our economies did not involve any reductions in wages; Mr. Towl had agreed that Cumberland would take care of any employees we might not require. One girl in our office, with a comptometer, took over the work which was reported to have required much of the time of fifteen men at Oil City, Pennsylvania, who had been receiving the crude oil runs by telegraph daily from each Kentucky lease and had been computing the number of barrels by hand. That was 25 years ago. Today, a large electronic computer could do the same work in a few minutes each day.

. . . . .

The purchase of the Cumberland Pipe Line . . . was the most significant acquisition we ever have made.

After a year of negotiation, Ashland purchased for $420,000 the pipeline, in which there had been a total investment of $3,600,000, and on September 26, 1931, organized a new company, the Ashland Oil & Transportation Company, to operate the line. The terms of the sale were $60,000 in cash and four $90,000 4-percent notes payable at six-month intervals. The down payment was less than the value of an inventory of crude oil included in the purchase but not shown on Cumberland's books. The discrepancy (more than $75,000) resulted from a 3 percent evaporation allowance for crude oil received from the producer. The evaporation had possibly been as high as 3 percent at one time, but over the years the New York management of Cumberland had failed to ascertain the actual amount of crude oil held.

The saving in transportation cost soon after the acquisition was approximately equal to the installment payments. In effect, Ashland acquired the pipeline at no investment.

The Cumberland purchase also gave Ashland several experienced executives, including W. H. Keffer, a future vice president. Most important, it gave the company security in transportation of its raw material supply. The purchase also permitted better planning for the continuous flow of oil from wells through the refinery.

The purchases of Tri-State Refining Company in 1930 and of Cumberland in 1931 were the two major acquisitions during this early period of expansion. They set the pattern by which Ashland continued to grow in later years: emphasis was on purchase of old equipment which could not be operated profitably by former owners; immediate and detailed attention was given to economies and improvements so that operations could quickly become profitable; acquisitions were balanced with previously owned equipment; fixed charges were held to a minimum through astute bargaining, by which property was purchased at a fraction of its original cost; and expansion generally occurred at times when the oil industry was depressed.

EFFECT OF GOVERNMENTAL FACTORS ON COMPANY POLICIES

Changes in the philosophy of the new federal administration in 1933 had their effects on the company. When it appeared that there was a possibility that the refinery employees would be unionized, the management took the initiative and actively encouraged organization. Blazer preferred an industrial type of union because he wanted to have the employees in one association. He wanted to know what he was confronted with and to forestall possible future organization by craft unions which might be more unfriendly to management. The result was that the company invited organization by the predecessor of the Oil Workers International (CIO).

The NRA code with its wage regulations caused a change in company policy. Blazer wrote to the Swiss board in August, 1933:

We are trying to get our plant in exceptionally good condition so that we will be able to reduce the size of our organization a little later. If our plant is unionized, which now appears probable, we will have every incentive to keep the size of our organization at a minimum figure, even to the extent of adding to our equipment expense. . . . With relatively cheap labor, it has always been our policy to keep our plant investment at a minimum figure even though it meant, in many cases, extra labor expense. To a certain extent that policy will now be changed. Our hourly labor rates in the plant will now average 40 per cent higher than heretofore.

66

At this same time, the oil industry faced the question of the desirability of price fixing by the government. Blazer resisted such efforts, although he felt that most executives within the industry seemed to favor price regulation. In a letter to Combs in 1933 he wrote:

A very large section of the oil industry wants price regulation, but I have been afraid of any form of artificial price-fixing, since under government regulations we might lose our natural economic advantage. Given a free competitive market, I feel that with a better organization and better equipment than ever before, we will be able to successfully cope with whatever problems may arise, but with the uncertainties and possible favoritism of artificial regulation, we might find ourselves in a less favorable position than in the past.

During the decade after 1928, Blazer became confident that his small, compact organization had advantages which would enable it to maintain profitable operations if competition were allowed to operate. Then came the death of Thomas A. Combs, and at the moment when Blazer had consolidated his approach to the operational problems of a small refinery, he found himself in charge of a medium-size, partially integrated oil company.

4

RELUCTANT ENLARGEMENT

Ashland Oil—A Medium-Sized Company

ASHLAND OIL & Refining Company was the consequence of the legal consolidation of Swiss Oil Corporation and Ashland Refining Company on October 31, 1936. Several factors made this consolidation desirable; however, the proximate cause of the amalgamation was a Kentucky law passed in 1936 which would have increased the taxes of the two companies if they had remained separate.

Although not stated in print at the time, the death of Thomas A. Combs in 1935 should be included as a reason for the consolidation. Combs had lived in Lexington while acting as president of Swiss Oil Corporation from 1922 until his death. His influence had been of such proportions that it would be reasonable to assume that no relocation of the home office would have been considered while Combs was the president. Yet the change in the relative position of the two companies and the continued increase of Blazer's prestige and power had moved the center of influence from Lexington to Ashland. Consolidation, therefore, furthered the potential coordination between the production and other phases of operations through the central location of offices.

The financial details of the consolidation were quite simple.

All of the 8-percent preferred stock of Ashland Refining Company had been called under the provisions of the stock agreement, and the 10-percent minority interest owned by Blazer was bought by the company at its stated book value. The total capitalization of the new company was $2,000,000; it included 1,000,000 shares of common stock, exchanged share for share for the common stock of Swiss Oil Corporation, and 100,000 shares of 5-percent preferred stock, half of which was offered to the public at the time of consolidation. These securities were traded on the New York Curb Exchange.

Blazer was interested in obtaining publicity from the Exchange listing in order to lay the foundation for further financing. Up to this time the stock of Swiss Oil Corporation had been traded publicly on the New York Curb Exchange without formal listing by the company. After the consolidation, Blazer gave special attention to the development of a good name in investment circles. The results of this attention were important to the future growth of the company; the way was being paved for further expansion should the opportunity arise.

EXPLORATION AND DEVELOPMENT

Blazer's influence on the company's operations led to emphasis upon refining and integration forward into marketing. Yet the extent to which the refining operations could be expanded was limited by the supply of crude oil available in the eastern Kentucky fields. The discovery of new fields in Illinois in 1937 was a fortuitous event that was prerequisite for expanded refining operations at the Catlettsburg location. These fields also gave Ashland Oil a suitable area in which it could expand its own producing activities. Until these fields were discovered, Ashland Oil had no undeveloped properties of importance.

Until 1942 Ashland's management either purchased its crude oil or bought marginal properties. It saw no reason to extend operations to any great degree into exploration and development. Blazer reasoned that in spite of the numerous problems which confronted the nonintegrated refiner, he could maximize profits from the use of his limited funds by increasing operations in branches other than production. However, after the Illinois fields became important potential sources of crude oil, the company obtained leases in western Kentucky and a separate exploration department was organized. This new effort represented one of the few times that the company entered a new activity without first purchasing a small marginal company around which operations could be built.

A small refining company initially entering exploration meets numerous problems. The extensive areas available for drilling tend to require geographical departmentation. The necessary capital funds discourage large operations. The fact that leases to the choice areas have been secured by other companies forces the newcomer to drill a large percentage of wells in unknown areas. The very nature of the exploratory operations requires persons of different skills and attitudes from those in refining and marketing. Many companies find that these operations can best be performed if the operations are completely separated in a different corporation.

As a means of averaging out some of the risks of drilling, Ashland's policy of developing campaigns jointly with independent drilling specialists greatly increased the crude supply for its refineries. Such an arrangement with producing specialists also enabled the company to gain the benefit of experienced personnel and an organization that was adapted to the problems of exploration.

In the early part of World War II, independent producers who were aided financially through loans by Ashland Oil brought in important new production. The company organized offices in Henderson, Kentucky, and Grayville, Illinois; later, in 1943, it opened an office in Tulsa, Oklahoma. A vice president directed the exploration operations. In each office there were at least a landman,[1] a geologist, and an engineer. However, company-owned wells outside the state of Kentucky did not begin to produce oil until the fiscal year of 1943.

MODERNIZATION AND EXPANSION OF REFINING EQUIPMENT

Coincident with the development of the new fields in Illinois—a prerequisite for any enlargement of the refining throughput—was the innovation of catalytic cracking. Blazer reacted to this drastic technological change by the following comment to the board in July, 1938:

The reason that I prefer not to increase our cracking capacity at this time is that I believe there will be revolutionary developments in the arts of cracking within the next year or two. Thus the additions which we are making at this time should be thought of as being somewhat temporary and only for the purpose of carrying us over until certain new processes are more fully perfected and made available to licensees. . . . As you know, until this year we had neglected our refinery in the matter of buying new equipment. Aside from buying new storage tanks, pumps, pipe lines, and asphalt-blending equipment, we have done no construction of importance for practically twelve years. The shell stills which will be abandoned at the end of this year were built in 1922; the cracking plant was built in 1926, but appears to be still good for a number of years of profitable operations. The shell stills were originally intended to handle 1,000 barrels per day of crude oil. For some years we have been getting through them 4,000 to 4,500 barrels. The cracking plant was designed to handle 750 barrels per day. We are running from 2,300 to 2,400 barrels. We are fortunate in having been able to hold off our modernization plan for so many years and still maintain reasonably efficient operations and good earnings.

The industry had experienced a relatively prosperous year in 1936, but the later 1930's were a time of recession. In spite of these conditions and the desire to postpone major capital

[1] A landman considers the purchase of new properties and negotiates agreements.

expenditures, Blazer began a modernization program in July, 1938. A 7,000-barrel combination atmospheric-vacuum distillation unit was ordered at a cost of $222,000. Since the contractor had relatively little business at the time, he gave special attention to this comparatively small job, and the unit was placed in operation in March, 1939. A small polymerization plant was added soon thereafter. Blazer reported the company's progress to the board:

As you doubtless know, the greater part of the capital expenditures, as listed, is associated with our program of almost doubling our refining and transportation facilities. Last May we refined 130,000 barrels of crude. This month we expect to refine between 235,000 and 240,000 barrels with another 20,000 available in July.

In 1939 the octane race started in its full intensity. The new Houdry catalytic cracking process promised to accelerate new changes. All companies felt the pressure to improve their gasoline quality. Blazer wrote the board in January, 1940:

With all major companies improving their gasoline within the last 60 days to meet the quality of the new Houdry gasoline, the burden upon us, as refiner of Eastern Kentucky crude is so serious as to be almost unbearable. We are adding excessive lead and benzol.

As part answer to this octane race, the old Dubbs thermal cracking unit was remodeled, increasing its capacity to 9,000 barrels per day. These improvements not only increased capacity but achieved better balance, decreased the number of shutdowns, and improved the quality of gasoline. In addition, the management increased its capacity for high-grade gasoline by adding a reforming unit capable of 1,500 barrels per day. The old Tri-State refinery was dismantled and used in the new construction at Leach, near Catlettsburg. This utilization of old parts decreased the costs of new construction and kept overhead low.

By 1940 the capacity of the refinery was 10,000 barrels per day; by 1942 it had increased to 18,000. This modernization program, immediately prior to World War II, had an important effect on

72

future growth, since the operation of a large catalytic cracking unit for the account of the government depended upon this expansion. Blazer explained these developments in a letter on April 14, 1942:

> There are comparatively few refineries situated as we are that are able to run at full capacity, having adequate supplies of crude oil and a ready market for their output. Our unusual position results principally because of our water transportation facilities, although we are dependent to some extent on tank cars.
>
> . . . . .
>
> It appears that under war conditions the smaller oil companies will suffer more than the large ones. For example, if we were only slightly smaller, we would not be in a position to support the proposed new government-financed plant. The program we are entering into practically requires that we refine a minimum of 15,000 barrels of crude daily. . . . I am afraid we will be forced to become a larger company whether we want to or not. . . . I am afraid that in order to protect what we now have, it will be necessary for us to continue to grow.

In numerous other letters Blazer had referred to his desire to become only as large as technology required. He appeared to feel that the well-managed smaller company had many advantages in the industry and that the trouble experienced by other small companies was primarily the result of inefficient operation.

Immediately after the entrance of the United States into World War II, Blazer spent a large part of his time in Washington helping to fit independent refining companies into the war program. The Secretary of the Interior appointed him chairman of the Refining Committee for District Two of the Petroleum Administration for War. Blazer explained to the board at the time: "Due to disturbing conditions resulting from the war, most oil companies are now being run from Washington. I find it necessary to be there most of the time." During his absence the subordinate executives had their first chance to work without the close supervision of their chief executive. Blazer once observed after returning from a trip: "I returned to Ashland

73

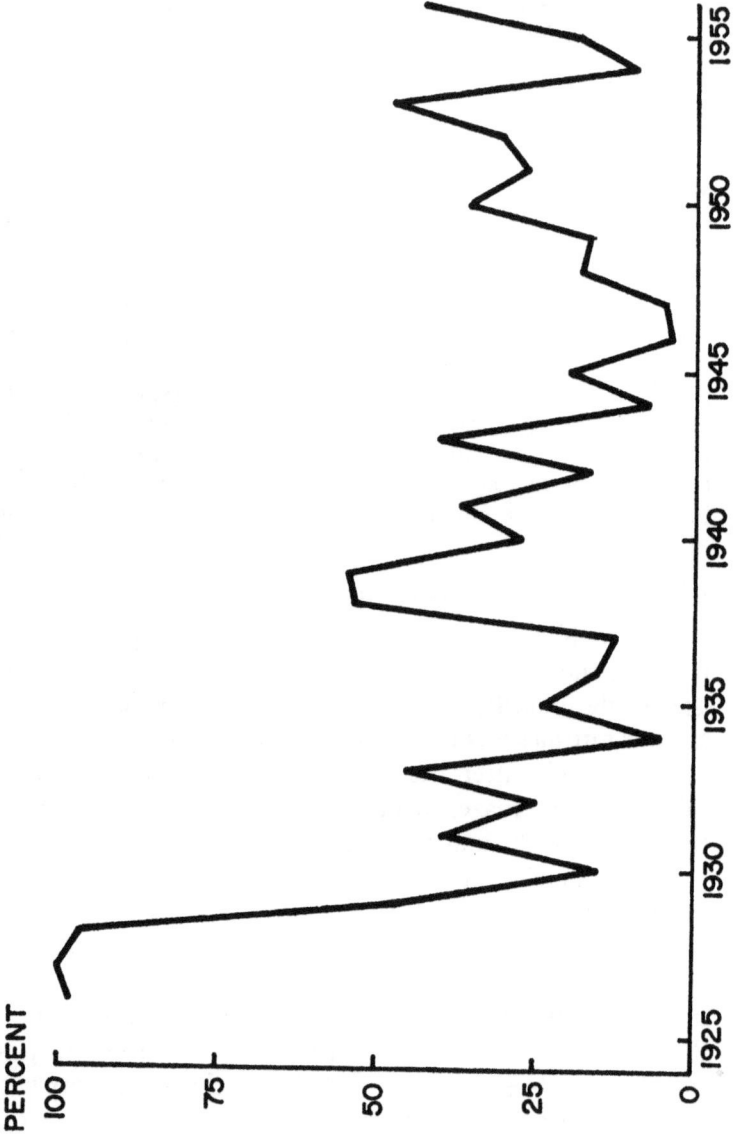

EXHIBIT 13—PERCENTAGE OF TOTAL CAPITAL EXPENDITURES PLACED INTO REFINING OPERATIONS, ASHLAND OIL & REFINING COMPANY, 1924-1956

74

today after being away for ten days, and I am glad to report that our organization functions smoothly with a minimum of supervision from me."

The final major expansion in refining during the wartime period resulted from a contract, dated May 5, 1942, with the Defense Plant Corporation. As a result of this agreement, a thermofor catalytic cracking unit for manufacturing 100-octane gasoline was built on a sixty-acre plot adjacent to the Leach refinery. Ashland operated it for the account of the government. In order to obtain this plant, Ashland committed itself to capital expenditures of $250,000 to adjust the Leach refinery operations to supply feed stock for the new catalytic cracker. During 1943, 40 percent of Ashland's capital expenditures went into refining. In addition, the government spent more than $16,000,000 in the construction of the plant. The unit went on stream July 1, 1944, using approximately 9,000 barrels per day of the throughput of the company-owned refinery as feed stock.

During 1939-1944, as a result of the war effort, the management had expanded from a small refinery of 1926 vintage processing 4,500 barrels per day to a large modern plant with a capacity of 22,000 barrels per day. Increase in capacity was only part of the expansion of the organization, for by 1945 the Ashland management was in charge of the operation of the new catalytic cracking plant which produced 100-octane gasoline for the armed forces.

The story of these years of expansion is one of a continual shifting of emphasis from one branch to another in response to changes in the basic industry conditions. Exhibits 7, 13, 16, and 17 show the fluctuation in the relative importance of capital expenditures in the four branches from 1924 to 1957. Marketing received a large percentage of capital funds during the 1930's as a result of the intense competition of those years. Capital expenditures in refining depended primarily upon the technological developments in the industry. These developments were not periodic, and relatively large sums were required at times when improvements were made. Transportation received an increased percentage after 1937, because it became necessary to transport crude oil from more distant fields.

For the first fourteen years, the refining company had operated with such a small staff of office personnel that the entire home office was located on a single floor of a bank building in Ashland. On January 15, 1938, the offices were relocated in a six-story building purchased and reconditioned in the previous year. The original cost of the building was low, and the rents from those portions not used by the company paid the operating costs for the entire building.

By 1938 Blazer gradually found additional departmentation desirable. Few functional departments (staff specialists) had appeared prior to the consolidation in 1936. With the accelerated expansion of operations, Blazer gave more attention to his organization; however, he continued to deemphasize the clear-cut separation of duties. In 1938 a functional personnel department was organized, and in 1939 exploration and development became a major operating department.

Managerial responsibility and prestige in the industry increased at an even more rapid rate than company sales. The expanded company-owned facilities and the large government plant placed new problems on the staff of subordinate executives, which was held to a minimum number by the wartime shortage of personnel. The experience obtained by these men enabled them to become adjusted to larger scale operations. The stabilization of the industry by governmental control permitted Blazer to delegate greater authority to subordinates. Many who had previously served in an assistant capacity became functional specialists.

76

The rapid growth of executive responsibility increased the need for another top executive with broad training. In order to meet this need and to provide a possible successor, Blazer obtained the services of a relatively young petroleum lawyer, J. Howard Marshall, who had served with the Petroleum Administration for War. Marshall, who was thirty-nine years of age (fifteen years younger than Blazer), had previous experience in academic administration, private legal practice, and governmental administration. As general counsel for the Petroleum Administration for War, he had become well known in the petroleum industry. He would be especially helpful in dealing with the government, since he "knew his way around Washington." Marshall assumed the presidency of Ashland Oil in 1944, a position he held until 1951, when he left the company. Blazer had continued as chief executive with the title of chairman of the board.

POSTWAR TRANSITION IN REFINING

Immediately at the end of the war, the government closed its plant; shortly thereafter, however, Blazer decided to continue its operation on a temporary basis. The annual report for 1946 explained the reason for reopening the plant:

While our organization of skilled operators was still intact, it appeared advisable to give the plant a fair test of its utility for the manufacture of high octane automotive gasoline. This operation served a dual purpose. It assisted the government and the company in ascertaining the value of the plant for commercial use, as well as in determining the extent of alterations required for the most economical peace-time operation. Although the six-month trial run under varied operating conditions, resulted in a financial loss to the company of several hundred thousand dollars, it has facilitated a realistic appraisal by the company and the government of the use-value of the plant.

After a prolonged period of adjustment and after government attempts to secure other bidders, Ashland Oil purchased the

100-octane aviation gasoline plant on December 11, 1947, for $2,150,000. The company agreed that it would make improvements costing $650,000. Negotiations preceding this purchase emphasized the great advantage that the company had obtained when the government first located the plant near Leach. No one in the petroleum industry knew the value of the plant as well as did Blazer; the properties were adjacent to the Ashland Oil refinery; another company which might have purchased the plant would have realized major obstacles in crude oil supply and marketing organization. The desire on the part of the Department of Justice to sell many such plants to independents gave an added advantage to Ashland Oil. On the other hand, Blazer knew that if another company were to purchase the plant, the future development of the Ashland Oil organization would be most difficult. He recognized that his refinery would need catalytic cracking facilities in order to maintain its competitive position after the war.

With this in mind, Blazer was able to arrive at a price which he thought would be favorable to the company but would be high enough to buy the plant from the government. He knew what other companies had successfully bid for similar plants. The assumption that no other company was actively considering the purchase was supported by the failure of others to inspect the facilities. In spite of the fact that the government made special efforts to obtain other bids, Ashland's was the only one made.

The purchase of the government-constructed plant again enabled the Ashland Oil management to keep pace with improvements in the industry without undue financial strain. The terms of $800,000 in cash and the remainder in 4-percent ten-year notes secured by a purchase money mortgage were adapted to the needs of the small company. The ownership of the plant not only gave the company sufficient facilities for production of high-octane gasolines, but provided it with enough high-octane gasoline to use as blending stock for less modern plants that it

78

might purchase in the future. Referring to this significant action, Blazer admitted to the members of the board: "The entire program, including the construction and acquisition of the aviation gasoline plant, involved many possible pitfalls, and looking backward, I am convinced that we got by some of them more through good luck than through good management."

During the negotiations for the "Number 2" refinery, as the former government plant was called, Blazer found an opportunity to lease and operate another refinery. This time the location at Niles, Ohio, was not on a navigable river but on the northern edge of Ashland's marketing area. In 1947 the company contracted with Western Reserve Refining Company to lease the refinery, and later it exercised an option to purchase it. Although this refinery was small and obsolete in design, it was able to make large profits during periods of good demand for fuel oil. Its relatively low cost of operation resulted from high morale of its workers, location near large steel mills which used heavy fuel oil, and the knowledge of the Ashland Oil management for obtaining the greatest amount of production from old equipment. These factors made possible intermittent profitable operations there during the next five years.

Although the refining units continued to be concentrated near Catlettsburg, the company had become a multirefinery operation through the addition of two small and obsolete plants —the one at Niles, Ohio, and one at Pryse, Kentucky. The Pryse refinery had been built in 1917 by Oleum Refining Company and purchased by Great Southern Refining Company at a time when Blazer was in its employment. It was connected to a pipeline which Great Southern had constructed from the producing properties in the Big Sinking district. The refinery, along with important affiliated producing properties, was purchased by the Texas Company in 1926. During World War II it had been operated at a rate of 1,600 barrels per day, but it had such high costs that it was extremely marginal. In 1945 Texas sold it to Ashland Oil along with the Kentucky producing properties.

Blazer had not planned to operate the Pryse refinery acquired by Ashland Oil in the package transaction, but it was extremely profitable during the ensuing two years. Several factors made possible its efficient operation until 1947: the experience of Ashland Oil in handling old refineries enabled the personnel of the company to decrease costs and increase throughput; the morale of the workers and their desire to make their jobs last as long as possible increased labor productivity to such an extent as to offset the disadvantage of the old equipment—during the period of refinery strikes in 1945, the Pryse plant continued to operate at capacity; and a shortage of fuel oil made it possible to obtain high realizations from the refined products. The refinery continued to show excellent profits up to the time it was dismantled. The management reasoned, however, that the Leach refinery could operate at more efficient levels if it increased throughput to supply the products normally produced by the Pryse refinery. Parts of the Pryse plant were of use in expansion of the throughput of the Leach refinery; storage and pumping facilities at the Pryse location continued as a part of the crude oil gathering system.

Throughout the war period Blazer planned for a quick adjustment to peace. Funds had been secured on favorable terms from a life insurance company during the war in order to finance the necessary expansion. In August, 1945, these loans were refunded into twenty-year 3-percent sinking fund debenture bonds. The company had retired its 5-percent preferred stock in April, 1945, and issued shares of 4¼-percent cumulative convertible preferred stock. In both cases, additional funds were obtained for working capital. This entrance into the securities market not only increased the amount of working capital but introduced the company to many investors on the national market for the first time. The financial policy was to maintain a strong cash position while building the potential for securing additional funds in the future.

Blazer saw that postwar refining facilities would be greatly different from prewar equipment. The government-built plant furnished the company with catalytic cracking equipment and workers trained in its operation. Ashland Oil had prepared for a continuation of the "octane race."

Blazer also gave special attention to the potential effect of the war on the company's geographical advantage. The "Little and Big Inch" pipelines might change the economics of transportation in the area in which Ashland Oil operated. Blazer took the initiative in promoting the use of those pipelines for the transportation of natural gas in place of crude oil or refined products. This initiative was expressed in the form of testimony in congressional hearings and in the acquisition by Ashland Oil of promotional stock in the new venture. The localized nature of Ashland's marketing territory made the company vulnerable to a decrease in competitors' transportation costs into the area.

Toward the end of the war it was clear that in order to move the much larger throughput of gasoline, it would be necessary in the postwar period to inaugurate an aggressive sales program on an expanded scale. The company made preparations during 1945 for the program that it carried out during 1946.

The annual report for the year 1945 explained the situation:

Our share of the market demand for gasoline has greatly increased as compared with the pre-war period. This has been accomplished in part by obtaining many additional distributors in the area where we have marketed for many years, and also by using our low-cost water transportation facilities to increase the radius within which we can ship profitably. It is believed, however, that until new cars are on the roads in much larger numbers, gasoline consumption will continue to be disappointing and that we will not have sufficient business to permit the most profitable and efficient use of our refinery and transportation facilities which were considerably enlarged during the war.   .   .   .   .   .

It is believed that our company, heretofore engaged principally in refining, transportation and distribution, is now also firmly established in the producing branch of the petroleum industry.
        .   .   .   .   .

81

BLAZER AND ASHLAND OIL

We believe that we have been successful in attaining our objective, as stated in last year's report, of having the company in a strong financial position at the close of the war, with investments in transportation, refining, and marketing facilities sufficiently charged off to cushion the necessary postwar adjustments.

ADJUSTMENTS TO POSTWAR MARKET FLUCTUATIONS

Immediately after the end of World War II, the industry generally foresaw a period of weak demand for oil products until the automobile companies could produce a large number of cars and consumers could convert furnaces to oil. In 1946 the sales outlook for Ashland Oil was not especially bright. By the summer of 1947, however, executives of oil companies could foresee that oil products were going to be in short supply, especially domestic fuel oil.

The management of Ashland Oil was able in 1947 to expand its manufacture of those products in shortest supply by operation of its small, relatively obsolete plants and by use of its refining flexibility. Crude oil was obtained by its barges from areas not serviced by pipelines. The small, centrally controlled organization was able quickly to adjust its operations to increased throughput. The emphasis on flexibility which had been indoctrinated into each member of the organization paid off handsomely.

In 1947 and early 1948 the supply of crude oil was tight. A company which had only slight integration in production was on the spot. Yet contrary to the usual generalization by observers in the petroleum industry, Ashland Oil was able to obtain ample supplies of crude even in this period of exceptionally short supply. In order to understand this success, it is necessary to understand Blazer's efforts to develop a crude oil supply during the preceding ten years.

The solution to the problem of crude oil supply for a growing refining company can be through the ownership of crude oil gathering pipelines, purchase of oil from independent producers in the crude oil market, integration backward through the purchase of producing properties, and a long-range program of exploration, development, and production.

Purchase of oil involves either the problem of insuring sufficient crude oil to permit capacity operations of pipelines and refineries in times of relatively short supply, as happened in 1947-1948, or the problem of being caught in a refiner's "price squeeze" in times of plentiful crude supply, as happened in 1948-1949. Prior to the era of prorationing (before 1935)[2] these risks were counterbalanced by the maintenance of close working arrangements with independent producers and by the ability to take advantage of low-priced crude in times of overproduction. After 1935, other managements observed that prorationing laws make it advantageous for a refiner to integrate backward into production. In times of weakness in the prices of refined products, the refiner who has been dependent on the crude oil market faces a price which lags in any decline. Since the beginning of prorationing, prices have become more stabilized as a result of regulatory actions limiting crude production. With relatively constant crude oil cost (approximately three-fourths of the total cost of a refiner's throughput), any decline in the price of refined products has caused a serious decline in refining margins and has made stable, profitable operations difficult for the nonintegrated refiner.

In 1938 Blazer felt that a primary problem was to secure an adequate supply of crude for his refining operations. Profits from the producing operations themselves were only secondary. The new Illinois fields encouraged the Ashland Oil management to plan refinery improvements based upon the assumption that adequate crude oil would be available from independent producers.

In order to expand its supply, the company looked to the

---

[2] A group of laws was passed for the purpose of conserving crude oil supplies and of protecting the property rights of owners in underground reservoirs.

improvement of transportation facilities through both pipelines and towboats. In the first stages of expansion in the Illinois fields, the management emphasized increasing gross rather than net production.[3] It obtained its greatest security through the ownership of gathering pipelines and the collection of oil at the wellhead.

No refiner can be certain of purchasing all of the oil that he needs simply by paying the market price. Price adjustments are imperfect; price does not perform its rationing function completely. The crude oil market is based upon a posted price set by an important buyer in a particular field. This buyer usually states that he will accept oil delivered to him at the posted price. The seller is interested in getting the best price that he can for his crude oil over a period of time without wasting time haggling over each transaction. Nonprice factors are important in these relationships. Therefore, the nonintegrated refiner often depends on his reputation with the sellers in the field. In times of short supply, he may find it necessary to pay a premium in order to secure sufficient crude. In times of oversupply, he attempts to take all the production of the wells even though there is no legal compulsion; he tries to keep his connections for protection in the event of another period of short supply.

Although Ashland Oil owned little net production and tried to shift its purchases according to relative price differences, under such environmental conditions Blazer found that the crude oil supply remained inflexible. Integration might have contributed to greater flexibility through closer control over volume of production at the wells.

The fact that a major percentage of the supply of crude oil has been owned by integrated companies has minimized the amount available to the nonintegrated refiner. Moreover, the integrated operator's interest might be more closely related to

---

[3] *Net production* refers to oil produced for the account of the company only, eliminating any part in which others have interests. *Gross production* refers to all oil produced by wells in which the company has any interest.

the interests of the seller of crude oil than to those of the non-integrated buyers. Since the integrated company stands to gain as a producer from price increases of crude oil, the nonintegrated refiner finds himself with few other oil interests fighting an increase in crude oil prices.

In the light of these problems of crude oil purchasing, Ashland Oil from time to time integrated backward through purchases of producing properties. Most acquisitions were "stripper" or marginal wells which produced such a small quantity of crude oil per day that their out-of-pocket costs approached their gross receipts. The price of crude oil, under such conditions, has a direct effect on whether the wells are worked.

The refiner who owns a stripper well is closer to the cost situation of the refiner who must purchase crude oil on the market than he is to that of the owner of a well producing at higher volume rates. After a flush well starts to produce oil, the out-of-pocket expenses of lifting it to the lease tanks are a fraction of the price of the crude oil. In the short run, the owner of these larger wells finds his out-of-pocket costs per barrel a mere fraction of those of the operators of stripper wells.

The amount of attention necessary to operate a large number of marginal wells makes them unattractive to large producers. Ashland Oil expanded its ownership of small wells in the Illinois basin in order to operate its gathering facilities at a more profitable volume. Properties became available at reasonable prices from time to time and offered a means of securing an increased supply of crude. The development of newer methods of secondary recovery, by water flooding, has made this program more attractive in recent years. Ashland Oil has gained valuable experience in the unique aspects of profitably operating such marginal properties, which have a potential value for secondary recovery.

One example of Ashland's purchase of a group of stripper wells was in March, 1945. After brief negotiations with the management of the Texas Company, Blazer was able to buy all of that company's oil producing properties in eastern Kentucky. The primary objective was to secure four hundred wells in the Big Sinking district, located near extensive holdings of Ashland Oil and serviced by an Ashland Oil pipeline. In addition, as previously mentioned, a small refinery at Pryse was included in the transaction.

An intensified program of exploration, development, and production was undertaken by Ashland Oil influenced by the excess profits tax during World War II. Intangible drilling costs which could be charged as expense decreased the effective cost of finding new crude oil. The 1944 annual report explained:

Your company's principal expansion during the past year has been in its crude oil producing department and especially in exploratory drilling. Expenditures for leases, drilling and the equipment of new wells amounted to $1,655,630.76 of which $509,734.73 was charged against earnings. Of the 146 wells drilled during the year, 99 proved to be productive. That is a more favorable percentage of producing wells than ordinarily would be expected. Oil production from wells, wholly or partially owned by the company, increased 176 per cent during the year from 2,122 barrels per day to 5,865 per day. An even larger drilling program is planned for the coming year. There is exceedingly keen competition, however, for favorable locations at which to drill. Many companies having large earnings subject to excess profits taxes are inclined to take greater risks than usual in the drilling of "wildcat" wells. Funds, that would otherwise be paid in taxes, are being spent to discover new oil fields. The spending of such funds for this purpose appears to be in accord with government policy to assure adequate supplies of oil both for the war effort and future civilian requirements.

Expanded exploration activities continued through 1948. Blazer's observations to the board make it clear that he recognized at this time the advantages of greater crude oil production:

We are more inclined to be influenced by the fact that we are always potentially short of crude oil supply for our operations, although by making strenuous efforts our potential shortage never catches up with us. For example, our anticipated refinery throughput of 27,000 barrels per day for the month of June is almost exactly in balance with our anticipated available purchases of crude oil. They are within one per cent of each other.

The company's earnings indicate that it had developed management techniques capable of meeting this unusual industrywide shortage of crude oil. In spite of success in meeting the problems of a partly integrated company, Blazer looked forward in a monthly letter to a time of less urgent problems of coordination: "Eventually we expect our production to be our most profitable department, since it represents a net investment of more than four and one-half million dollars out of a total of eight million."

An example of the type of difficulty faced by a firm with very little company-owned crude oil production appears in another letter from Blazer:

Superior Oil Company which has been supplying us with approximately 3,000 barrels per day of crude oil, constituting our most important single source of supply, cut us off entirely last week because we were unwilling to go along with them in their frank effort to force an increase in the posted market for crude oil in the Illinois basin. . . . They gave our oil to Standard Oil Company (Ohio) without any financial inducement, this being handled in much the same way that they took 2,500 barrels of oil per day away from Aetna a year or so ago and gave it to us.

Exploration and development continued after the end of World War II, making possible an increase in gross production from 6,636 barrels per day in 1945 to 9,438 barrels per day in 1946. Net production increased in the same period from 3,197 barrels per day to 5,723 barrels per day. In spite of continued emphasis on this phase in the next two years, gross production increased only to 10,124 barrels per day by 1948. Shortage of steel and of good drilling areas prevented a greater increase.

In 1947 it was mandatory for Ashland Oil to obtain as much security of its crude supply as possible. The company emphasized increasing its gross production from oil wells in which it had only a partial working interest. The extra time and expense required to increase net production discouraged the company from stressing ownership of the total working interests. The management felt that the large amounts of funds needed in the producing branch required the spreading of risks of exploration. The annual report for 1948 explained the policy that had developed:

The average share owned by the company in this increased number of wells decreased from approximately 61% to 46%. This was in accordance with our policies of spreading our risks and developing an increasing number of independent-producer relationships.

. . . . . .

The large increase in available supplies of crude oil has been achieved principally by cooperating with independent producers and assisting them in finding, developing, equipping, operating, and financing producing properties.

Blazer illustrated again in a letter to the board that at such a time, integration in production appeared to be advisable: "Over the past few years, it has been our purpose to invest our large refinery and transportation earnings in crude oil exploration, recognizing that our crude oil supply has been the weakest point in our set-up."

Following this renewed interest in additional crude oil production, the company in 1947 looked to the Middle East for crude oil. Because of the large expenditures required and the risks involved, some of the independent oil firms organized American Independent Oil Company. Ashland Oil held 12.69 percent of the common stock. Broad powers were given the officers of American Independent, which searched in vain for a new field in its concession for six years before it finally discovered the Wafra field in the Kuwait-Saudi Arabia Neutral Zone in 1953. Several of the stockholder firms became dissatisfied with the management of the company, and litigation extending over a number of years resulted. Just prior to the discovery in 1953, Ashland Oil was asked to provide additional funds as a loan to

"Aminoil." The funds were advanced, but it was clear that Blazer was discouraged.

After the slow start, American Independent finally began to produce crude oil in profitable amounts—22,000 barrels per day just prior to the Suez emergency in 1956. The loans were ultimately paid off, and by 1957 it was evident that this venture would be successful. Ashland Oil by this time had begun to search in other foreign countries for crude oil: it joined a syndicate to explore in Venezuela and organized Ashland Guatemala Company to acquire leases in Guatamala.

COMPARISON WITH SOHIO'S EXPLORATION ACTIVITIES

The Standard Oil Company (Ohio) offers a useful comparison with Ashland Oil's producing activities. An important competitor, by 1950 it was only slightly larger as measured by refinery runs and sales. The Standard company had never owned any significant crude oil production; therefore, when it made its decision in 1942 to move into that branch, it was necessary to start at the very bottom. Its success in the other three branches had provided funds to support a major effort in production. The excess profits tax was also a primary reason for integrating backward.

The decision of Sohio's management to increase the amount of its net production was vigorously carried out. It developed exploration activities and purchased a number of small producing companies. After the outbreak of World War II, the small producers were especially willing to sell: crude oil prices were frozen; income tax rates were high, and a sale would make their receipts capital gains, thus lowering their taxes; the scarcity of supplies made it more complicated for the small producer to operate economically. In the opinion of the executive in charge of Sohio's producing operations at that time, the primary mistake during those years was that the company did not buy even more small producers while the opportunity was good. By 1953 Sohio's net production had increased to 23 percent of its refinery runs, while Ashland's had dropped to less than 10 percent.

Sohio's exploration plans extended over a five-year period. Its management reasoned that expanded exploration over short periods would not show sufficient returns to cover the large expenditures that were necessary. If a short-term viewpoint were taken, operations might be curtailed before their full returns had been realized. The cash income generated from past production proved insufficient to cover the new operations for the first five to six years of continued efforts. Only after such a long period could the realization from crude oil begin to cover the necessary large expenditures.

During World War II both Ashland Oil and Sohio turned to production of crude oil because of the effects of the excess profits tax. Ashland Oil started with some production; Sohio started with none. Sohio had sufficient financial resources to plunge heavily into production, but Ashland Oil had limited funds for production because it was growing rapidly in transportation and refining as well. Sohio committed a relatively constant amount of capital expenditures yearly to production; Ashland Oil adjusted its actions to changing situations.

DEEMPHASIS OF EXPLORATION AND DEVELOPMENT

Following the extreme shortages of 1947-1948, the oil industry experienced a sharp and rapid change in its prospects. By the middle of 1949 the market had so weakened for residual and heavy fuel oils that the problem had changed to one of too much crude oil; the excess profits tax was no longer a factor.

As a result of these changes, Blazer's 1949 letters to the board emphasized flexibility as more important than consistency with previously quoted comments:

It appears that these operations [crude production] for the past four years prior to this year have shown a net loss of more than a million dollars. . . . It is my impression that with the prospect of being able, for some time, to purchase from others all of the oil we need for refining requirements, there is little advantage to be derived from our present efforts to find crude oil production. Apparently we have not been able to put together a successful crude oil finding organization. I am confident that we can purchase crude oil reserves on a much more favorable basis than it has cost us to find them. . . . Our investment in that part of our business is so substantial that I believe we should not continue to expand in that direction in the absence of a profitable return. . . . With prospect of further shrinkage in refining and transportation profits, we won't have the "cushion" in the future that we have had in the past. We will be confronted with the necessity of making every department "stand on its own legs" yielding a fair return on the investment.

Following this policy, three western offices were consolidated into one. Blazer felt that if any phase did not yield returns comparable to the other branches, the only alternative was the reduction of activity in that branch. Integration of itself was not essential; profits must be relative to the funds invested in the separate stages.

Retrenchment in 1949 was based upon the assumption that the production department's profits should be computed for the same time period as other departments' profits. The idea was that the concept of flexibility and adaptability found to be important in the other branches could be applied in the same degree to the production branch.

The expansion in exploration in 1942-1948 was caused almost entirely by the effects of the excess profits tax. Reasons advanced by other managements supporting their integration into production may have been considered but were not of major significance in Blazer's decisions. Prospects for profits from crude oil did not appear as attractive to the Ashland Oil management as prospects in the other branches, especially in refining and transportation. Although the crude oil supply had been considered important by other managements, Blazer felt that he had never had trouble ultimately in securing enough crude oil even in times of extreme shortage (1948, for example). The fluctuations in refining margins continued to be difficult for the nonintegrated refiner, but Blazer felt that there were means of combating this problem over the long run.

Blazer considered flexibility the important counterbalancing force. He thought that the difficulties of the nonintegrated refiner could be met through several defensive measures: the refiner could shift purchases of crude oil to those grades more economically located for his operations; he could shift the refined products yield to those products whose prices were under the least downward pressure; he could pay closer attention to refining costs as a means of compensating for any unfavorable differential in the purchase price of crude. All of these measures placed a special emphasis upon a flexible organization, a flexible refinery, and a flexible transportation system.

With these ideas and the foundation in refining laid during the period 1936-1947, Blazer and Ashland Oil were on the threshold of a period of extremely rapid growth.

5

RAPID GROWTH BY MERGER

The Transformation to Large-Scale Operations

PROBABLY THE most astounding period in the history of Ashland Oil was the time of its rapid growth from 1948 to 1950. In this short period, as Exhibit 14 shows, the company's assets increased from $24,000,000 to $104,000,000. Although, as the previous discussion implies, Blazer had built a firm foundation for future growth, this expansion was not the result of a conscious program. In fact, in numerous comments through the years after 1938, he seemed to apologize for becoming larger. The 1947 annual report included one example of such a comment: "Growth, as such, is not an objective but because petroleum is an expanding industry, growth is almost unavoidable."

McLean and Haigh have stated in their study that the series of mergers by Ashland Oil was a part of a "major expansion program."[1] If by a program one means a formally organized emphasis upon expansion by purchase, the series of mergers did not result from such long-run rational planning. If one means that the management remained in a position to take advantage of a favorable proposition whenever it might occur, such a program existed throughout the history of the company. Many of

[1] McLean and Haigh, *The Growth of Integrated Oil Companies*, 641.

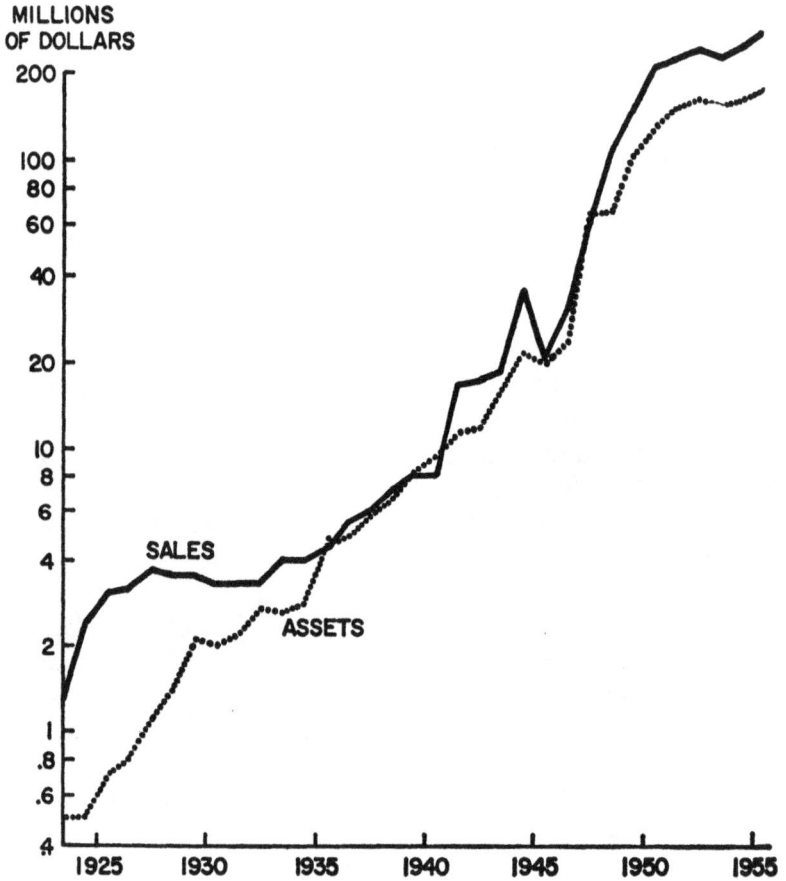

Exhibit 14
TOTAL ASSETS AND TOTAL SALES, ASHLAND REFINING COMPANY
AND ASHLAND OIL & REFINING COMPANY, 1924-1956

the mergers which did take place were in the stage of informal discussion for many years. Others were the result of a number of factors which suddenly directed the managements toward merger.

The impact of these mergers was so great on the development of the Ashland Oil organization that it will be most important to understand the background facts of each merger, the reason that each party desired consolidation, and the informal process used to achieve agreement. In the case of each of the four principal mergers which took place between 1948 and 1950, expansion in marketing of refined products was Blazer's primary objective. The fact that each of the companies owned a refinery at the time was of minor importance. The addition of these refineries, however, resulted in a fundamental change in Ashland's refining operations during the 1950's.

THE MERGER WITH ALLIED OIL COMPANY

The most spectacular of the mergers involved the agreement with the owners of Allied Oil Company of Cleveland, Ohio. Blazer had been in close contact with the owners and management of Allied from the beginning of both companies. After Allied had started operations from Cleveland as a fuel oil marketer in 1925, it secured some of its first fuel oil from Ashland's Leach refinery. Blazer kept in touch with Allied's owners, W. W. Vandeveer and F. R. Newman, by continued business and social contacts. This friendship was responsible for Allied's hiring Paul Blazer's nephew, Rexford S. Blazer, upon his graduation from college in 1928. Rex subsequently rose to the position of vice president and sales manager with Allied.

The first written indication that there was serious thought of a merger was in March, 1937. At that time one of Allied's owners was interested in a consolidation in order to expand the operations and to enable him to take some money out of the business, but the other owner was cool to the idea and indicated that he was too busy at the time to discuss the consolidation. Talks were renewed from time to time at petroleum meetings, but they did not become serious again until around 1944. Formal negotiations began in July, 1946. Allied was mainly a heavy fuel oil marketer; Ashland Oil was chiefly a refiner which produced heavy fuel oil. After 1940 Ashland Oil had been expanding its market territory to the upper part of the Ohio River, using towboats and barges as the basic means of transportation. A merger with Allied fitted in with Ashland's river system of transportation and connected with a market area served by the lake tanker fleet owned by Allied. In general the two companies seemed to be complementary.

Allied Oil Company had been organized on June 6, 1925, by Vandeveer and Newman. A temporary majority ownership was held by M. Bond, who supplied the additional necessary funds to start operations. The primary objective of the company was the buying and selling of industrial fuel oil in tank car lots. Its rapid rate of growth was chiefly the result of its specialization in the unique problems of supplying industrial firms with a commodity normally considered a byproduct by large integrated oil companies.

Promotion of this specialty business resulted in a type of contract which offered security of supply to the purchaser while making it possible for him to buy at as low a cost as the spot market would offer. Allied also developed efficient means of transporting and handling heavy industrial oil. It kept the highly viscous residual oil heated through much of its transportation to avoid the necessity of reheating at exchange points. The low value of this oil, its high viscosity, and the "dirty" characteristic of the equipment used for its transportation made it desirable to use water transportation and specialized handling equipment at terminals. Since residual oil could not be transported by pipeline, river and lake boats were the logical means.

After purchase of its first lake tanker and construction of a terminal at Cleveland in 1929, Allied increased its lake tanker fleet to six ships, with an aggregate capacity of 200,000 barrels. Demand for transportation services from other companies encouraged expansion into light petroleum products. By the time of the Ashland Oil merger, 90 percent of the oil carried was by contract for the account of other oil firms.

Allied used the Ohio River for barge movements. At the time of the merger, it marketed fuel oil in Ohio, Indiana, Illinois, Wisconsin, western Pennsylvania, western New York, and West Virginia.

In addition to its industrial fuel oil and transportation business, Allied marketed domestic fuel oil in Cleveland and Akron. To avoid a credit loss, the company entered the refining branch through purchase of a refinery at Canton. This plant, built in 1930 with a capacity of 1,800 barrels per day, operated until 1933 but remained closed for three years thereafter. Then it was leased for a year until the lessor failed; Allied was the primary creditor. Subsequently, Allied leased the refinery in December, 1937, and purchased it in April, 1940. Operations became very profitable as a result of the newly discovered wells in Illinois. By the time of the merger, the refinery had increased its crude oil capacity to 10,000 barrels per day and its thermal cracking capacity to 2,000 barrels.

Soon after beginning operations, Allied expanded its activities by purchasing crude oil from independent producers through Central Pipe Line Company, which gathered oil in the Illinois fields. At the time of the merger, Allied operated 170 miles of gathering lines.

As a means of supplying crude oil to its refinery, Allied extended its operations into the producing branch. Net production in 1947 was 567,191 barrels as compared with Ashland's 1,703,466 barrels; gross production was 1,131,274 barrels for Allied and 2,836,930 barrels for Ashland Oil.

Prior to the merger in August, 1948, Allied and Ashland Oil had developed two joint operations. In 1947 both companies

97

had participated in the organization of American Independent Oil Company, each owning more than 6 percent of the common stock. A second joint venture was Allied-Ashland Tankers, Inc., which owned two oceangoing tankers with total cargo capacity of 262,000 barrels of crude oil.

Several previous studies have pointed out a number of factors making merger desirable for Allied. Also, Vandeveer of Allied published his interpretation of the reasons for the merger. These publications emphasized the uncertainty of the financial status of the company in the event of the death of one owner. Since both owners of Allied had most of their wealth invested in the oil company, a serious problem of liquidity would have developed in the payment of a large inheritance tax. Because the stock of the firm was not listed on an organized exchange, there would have been a question as to the valuation which would be placed on the assets. Vandeveer summarized this viewpoint in a privately printed pamphlet:

There was no monopolistic grabbing of our facilities. We had done pretty well in competition with the largest oil companies in our marketing area and, given freedom, I have no fears for little business men in competition with their big competitors. No competitor bought us out to rid themselves of a troublesome rival. There was no threat of reprisal by any business man or corporation if we didn't sell. . . . The economic pressure—the economic necessity—was of government origin. . . . We would still be in business except for the unrealistic tax laws that forced us to sell our company.[2]

Butters, Lintner, and Cary concurred in this opinion: "In the absence of tax motivation to sell, the Allied owners clearly would have retained the ownership of their business beyond 1948. Quite apart from tax motivations, however, they might have decided to sell within the next decade or so, when they would have faced the necessity of retiring."[3]

[2] W. W. Vandeveer, *Mergeritis* (n.p., n.d.).
[3] J. Keith Butters, John Lintner, and William L. Cary, *Effects of Taxation on Corporate Mergers* (Boston, Harvard University Graduate School of Business Administration, 1951), 45.

Undoubtedly taxes were important in the decision of the owners of Allied to merge. But to find all of the fundamental reasons, one must look still further. One owner had been in favor of a merger with Ashland Oil in 1937, before the inheritance tax was so important. Furthermore, by 1948 large sums of money were required to modernize the Canton refinery. Allied had always had a serious problem of raising sufficient funds to continue its rapid expansion. After 1933 its current ratio was often less than two to one. It is not surprising, therefore, that an offer of $11,500,000 would be accepted by Allied's owners for an equity stated on Allied's books at $9,873,593. Additional advantages from association with a larger and more secure operation, potential appreciation from Ashland's stock, and diversification of holdings were reinforcing reasons to merge.

Once the Allied owners had decided to merge, there were many reasons why Ashland Oil should be a logical purchaser: the two managements had been in close contact since their organization; few larger companies were interested in Allied's type of business, while Allied's territory complemented Ashland's operations; the government's attitude from the standpoint of antitrust policy would be more favorable to a merger of Allied with a small concern such as Ashland Oil than with a larger company; and Ashland Oil was not so large that the Allied organization would be "lost"—the owners of Allied wanted to keep their organization intact.

The comparison of the development of Allied and Ashland Oil over the same period of years in an overlapping marketing territory yields several interesting differences between the companies. Allied was continually in "tight" financial condition in spite of large earnings; Ashland Oil placed an emphasis upon maintaining large working capital in order to add financial flexibility to its operations and expansions. Allied remained a completely closed corporation; Ashland Oil was publicly owned from its beginning and continued to seek additional funds from the capital markets, often even when it did not need them. Allied

ASHLAND | ALLIED

NET INCOME

| 66% | 34% |

STOCKHOLDERS' EQUITY

| 63% | 37% |

SALES

| 47% | 53% |

ASSETS

| 58% | 42% |

Source: Prospectus of Ashland Oil & Refining Company, August, 1948. Net income and sales are based on a pro forma statement for six months prior to March 31, 1948.

specialized in marketing; Ashland Oil specialized in refining. Allied based its sales program upon continuous relations with its customers by long-term contracts which might not yield a maximum return in the short run; Ashland Oil based its sales program upon developing market flexibility so that it might shift from one type of customer to another as conditions made it possible to increase profits from sales of the moment. Allied had used extreme departmentation in its later years; Ashland Oil had always emphasized the integrated character of its operations to such a degree that it was difficult to distinguish between branches. Allied's management had used a technique of staff meetings to gain coordination; Ashland Oil had used a bilateral method.

The merger of Allied was somewhat different from later mergers involving Ashland Oil. Allied was a successful company with large past profits, with high expectations for the future, and with no pressing operational reason for merging at the particular time (a period of prosperity in the oil industry). Moreover, Allied's size approached that of Ashland Oil, as Exhibit 15 illustrates.

The Allied-Ashland Oil merger on August 9, 1948, resulted in an expanded organization which was much larger than before. By this merger Ashland Oil suddenly faced problems of a large firm, thus requiring a change in its managerial techniques. The merger provided a number of new executives who had not grown up in the management of Ashland Oil. It offered an exceptional challenge to Blazer's managerial philosophy.

POSTWAR RECESSION IN REFINING

Soon after the Allied-Ashland Oil merger, the petroleum industry experienced a sudden change. In the fall of 1948 there was much uncertainty in the minds of executives of all companies as to whether the problems of the next six months would be ones of oversupply or undersupply. For example, in September, 1948, Phillips attempted to increase the posted price of crude by 35 cents per barrel. But the industry failed to follow the increase in price during the winter of 1948-1949, which turned out to be a relatively mild one. Companies found their inventories were too high, and some firms ran into financial difficulty.

During this sharp readjustment in the oil industry, Blazer considered numerous possible mergers. In his letter of January 28, 1949, he explained to the board:

Many of the smaller refineries are curtailing their operations because of reduced margins of profit and some of them are shutting down. As you know, numerous companies are attempting to sell to us or to merge. Prospects for most small refineries are not encouraging. . . . It is principally because of their shortages of cash that they are forced to make some kind of deal. . . . It appears, however, that at least for the next few months our Catlettsburg refinery will not be seriously affected and earnings in some of the departments of the company may improve.

The pressures on the Ashland Oil management during the fiscal year, October, 1948, to September, 1949, were exceedingly great. The entire industry was going through a sharp readjustment period. In addition, the Allied organization was in the process of being taken into the Ashland Oil operation. Blazer gradually was redesigning operational and managerial techniques to accommodate the consolidation. The personnel of Allied Oil Company were approximately of the same age and experience as were those of Ashland Oil, but by degrees many of the key executives were absorbed by the home office at Ashland. Blazer observed at this time in a letter:

I am working closely with Rex Blazer [Allied's new chief] in the operations of Allied and Captain [L. M.] Jonassen [chief of Allied's lake tanker fleet] is reporting to me on his tanker operations, although for all practical purposes, he and Rex are running their respective businesses. So far as their organization and their customers are concerned, we are taking no part in their operations.

102

In spite of these pressures, Blazer recognized that numerous opportunities were literally knocking at his door. Starting in January, 1949, much of the time of management was spent investigating companies which were interested in merging with Ashland Oil. It was at this time that serious study of Aetna Oil Company and Frontier Oil Refining Corporation was intensified. In addition, staff reports were made on at least three other companies, although no mergers resulted.

Probably the summer of 1949 provided challenges to Blazer's management techniques greater than any other time in the company's history. While other independent refineries were faced with critical situations, Ashland Oil continued to show large profits. The situation was summarized by Blazer in a letter to the board on June 29, 1949:

Gasoline prices appear to be holding steady. Heavy fuel oil prices probably will not go much lower but there may be some further reduction in the prices of light fuel oils and other distillates. We now have practically no production of heavy fuel oil at our Catlettsburg refinery and relatively small yields of distillates. Thus, keeping in mind that we are coming into the asphalt season, our return on the products from a barrel of crude oil at our Catlettsburg refinery appears to be well stabilized. Our profit last month of approximately 33c per barrel would be considered by most refiners a good profit under normal conditions. Conditions in the refining branch of the industry, however, are far less favorable than normal and I am told that most companies are realizing little or no earnings from their current refinery operations. . . . Earnings to be reported for the third quarter of the fiscal year will be disappointing. I anticipate they will show less than 70c per share on the common stock, giving a profit for the nine months period ending June 30 of about $4.00 per share. Assuming there is no change in the price of crude oil, I believe we should earn during the fourth quarter close to $1.00 per share, which would permit us to end our fiscal year with a net profit of about $8,750,000. If the price of crude oil should break, it might affect our current earnings, but over a period it would be advantageous to us. The relationship between crude oil prices and refined product prices is extremely unfavorable for refining; our refinery profits would be much lower except for our favorable trans-

103

portation setup. Since the amount of oil we are now refining is about six times as great as our net crude oil production, we are at a disadvantage in this kind of a market when refined products, selling in a free market, have declined sharply, whereas crude oil prices have been maintained by means of proration. Except for the pressure of our crude oil supplies we could cut our refinery crude oil runs to less than 30,000 barrels per day and continue to make almost as much gasoline with a larger profit per barrel of oil refined.

These profitable operations by Ashland Oil placed Blazer in an exceptionally strong bargaining position in discussions with other companies. The mergers following the one with Allied were therefore in a greatly different setting, even though they occurred within twenty months.

THE MERGER WITH AETNA OIL COMPANY

By the end of 1949, agreement was reached with the stockholders of Aetna Oil Company—45.53 percent of Aetna's stock was purchased with cash and the remainder was exchanged for common and preferred stock of Ashland Oil. Again the chief strength of the merged company was marketing; Aetna controlled a well-accepted brand of gasoline in Louisville and the adjoining areas. A notice to the stockholders of the annual meeting of January 16, 1950, explained the valuable assets of Aetna:

Aetna has specialized in the field of retail marketing of petroleum products under its own brand name. Its principal retail marketing territory embraces Central and Western Kentucky, Louisville, and Southern Indiana.

. . . . .

Distribution is effected largely through 39 bulk stations, from which 220 company-owned or leased service stations, some 580 dealers and approximately 2,000 consumer accounts are served. Where it does not have bulk plant representation in its territory, it has jobbers who purchase their requirements from Aetna and sell them in their markets under Aetna's brand name. Such jobbers, numbering 15 at the present time, serve 75 owned or leased service stations, 200 dealers and some 500 consumer accounts. Its direct and jobber distribution under its own brand name normally accounts for around 75% of its refinery gasoline output.

Using funds supplied by Commodore E. C. Benedict, James Duffy had organized Aetna Refining Company in 1917 and had built a refinery in Louisville. After Benedict's death, the company went into bankruptcy in July, 1921. Administrators of the Benedict estate purchased the company for the book value of the bonds and preferred stock and recapitalized it in the name of Louisa Company. In 1927 the name was changed to Aetna Oil Service, Inc.

As a result of a critical financial condition in January, 1932, the administrators of the Benedict estate gave operating control to Walter Abbott, a geologist, and Walton T. Davis, a chemical engineer. Abbott had first joined the company in 1923; Davis, in 1924. In 1936, by borrowing heavily, these two men bought controlling interest in the company. Throughout the decade of the 1930's Abbott and Davis carried the full weight of top and middle management. By 1939, after Davis had fallen ill because of overwork, the owners expanded the management group by adding three vice presidents: C. M. Alexander supervised refining and transportation; A. J. Brewer became the accountant; and J. B. Hutchinson directed sales activities. Abbott devoted most of his time to crude oil exploration and production in western Kentucky and Illinois, since the newly discovered Illinois fields offered an area in which the company could integrate into production. In May of 1941, Abbott died and Davis assumed the presidency of the company. By this time the refinery had a capacity of 8,000 barrels per day and Aetna's marketing territory had expanded to Cincinnati.

Soon after becoming president of the company, Davis volunteered for the armed forces. As a result, the operating management of the company during the wartime period was left to a committee of the three vice presidents. After the war, Davis returned to operating control and directed the company through the profitable period of 1947-1948. On June 26, 1948, however, the refinery suffered a serious fire which curtailed operations. The drop in prices of some of the fractions of crude oil in 1949 caused extra problems. Net income dropped from $661,903 in 1948 to $172,000 for the first eleven months of 1949.

Several factors made a merger with Aetna attractive to Blazer. Aetna's operations were complementary with those of Ashland Oil. Its refinery was accessible to Ashland's barge system, so that high-octane blending stocks could be transported from the Leach refinery to supplement the quality of Aetna's gasoline. Further, Ashland Oil could increase its intensity of integration forward by Aetna's company-owned sales outlets; few accounts overlapped.

Equally good reasons for merger existed for the owners of Aetna. Staggering capital outlays on the refinery were needed, and the problem of funding a new pension plan for employees would involve further large expenditures. The chief owner, Davis, had been under continued strain; his personal wishes were to lighten his load. As in the other mergers, tax was a factor. Net production of crude oil had never been over 15 percent of refinery runs; in the shortage after World War II, the pressure to secure sufficient crude oil was great. Other short-run problems were added by the serious refinery fire in 1948. Finally, the administrator of Abbott's estate had negotiated to sell the beneficiaries' interest to Ashland Oil; Davis faced the possibility that his 42 percent ownership would be a minority interest and he recognized that his position could be made more difficult in the event that his ideas differed from the strong management of the Ashland Oil organization.

THE MERGER WITH FREEDOM-VALVOLINE OIL COMPANY

At the time of the Aetna negotiations, another opportunity developed suddenly and was consummated within a period of a month. Freedom-Valvoline Oil Company was merged under the same financial arrangements developed for Aetna. It had been in the owners' minds for a period of years, but the fact that financing could be expedited if action were taken at the time of the Aetna merger encouraged all parties to negotiate without delay.

Final decisions relating to the Freedom-Valvoline merger were rapid. After contacting the owners by telephone on a Thursday afternoon in November, 1949, Blazer, accompanied by his controller, visited Pittsburgh on the next Saturday. The parties reached agreement on the following Tuesday. Total actual time of negotiations was not more than five days.

Freedom-Valvoline had had a long and colorful history. Studies sponsored by Ethyl Corporation have compiled many of the facts. The roots of Valvoline were in Continuous Oil Refining Company, an organization formed by Dr. John Ellis in 1866. Two years later the company's oil, recommended by George Corliss, the maker of the Corliss steam engine, was called "Valvoline." The trademark "Valvoline" was obtained in 1873, and in 1902 the company took the name of its nationally famous product.

The son of John Ellis remained active in the management of the company until 1930, at which time E. W. Edwards, a steel executive of Cincinnati, purchased control. A year later, Valvoline Oil Company purchased Galena Oil Corporation of Franklin, Pennsylvania, which had previously bought Signal Oil Company.

Freedom Oil Works Company in its own right had a history dating back to 1879, at which time a refinery was built. In 1889 J. W. Craig and A. J. Minke organized the company and participated in its management until 1912. Craig's nephew, Percy L. Craig, continued in active management until his death in 1929, when his son, Earle M. Craig, who was one of the owners at the time of the Ashland Oil merger, assumed managerial control. In 1938 part of the marketing facilities of the company were sold to Socony-Vacuum Oil Company. William G. Bechman bought part ownership in the company on January 1, 1939, and the name was changed to Freedom Oil Company. In June, 1944, Freedom Oil Company purchased Valvoline Oil Company and late in 1945 formed Freedom-Valvoline Oil Company.

At the time of the merger, Earle Craig and W. G. Bechman were the principal owners of the firm; Craig owned majority interest and was the chief executive officer. The company manufactured lubricating oils from Pennsylvania-grade crude gathered through a 1,400-mile pipeline system in western Pennsylvania, West Virginia, and southern Ohio, and through common carrier pipelines which terminated at the refinery on the Ohio River at Freedom, Pennsylvania. This refinery had a small thermal cracking unit, a vacuum unit, and a crude oil capacity of 4,000 barrels per day. Although operated principally for the manufacture of lubricating oils, it also produced a low-quality gasoline.

Until the time of this merger, Ashland Oil & Refining Company had specialized in a regional marketing area with no product being distributed nationally. Nor did it have a lubricating oil of its own manufacture. The importance of the merger to Ashland Oil was that it achieved a nationally advertised lubricating oil in the transaction. Facilities located on the Ohio River fitted in with servicing by the Ashland Oil barge system. The high-octane gasoline from the Leach refinery could be shipped economically for blending with the low-quality gasoline produced at Freedom.

Both owners of Freedom-Valvoline had entertained the idea of merging with another company; they had discussed it with several smaller companies but had taken no action. They had several reasons for their interest in merger with Ashland Oil. The stock of Freedom-Valvoline was not easily marketable, but Ashland's shares were listed on an organized exchange with a relatively broad market, a fact which would permit Craig and Bechman to diversify their investments, as they desired. The effect of high inheritance taxes contributed to the final decision.

Craig made a statement for the press at the time of the merger explaining:

> Higher and higher inheritance taxes, as well as income taxes, has meant that a business, developed by an individual or a small group of individuals, cannot be passed on to a second generation. It becomes necessary, in instances, during the lives of the principals who have developed the business to merge it with some larger company for whose securities there is a ready market. . . . Our representatives in Congress advocate help for small business itself, but they fail to realize that the tremendous taxes hurt small business more than big—and cause bigness.

The declining production of the Pennsylvania oilfields caused the cost of gathering crude oil to be very high. In addition, the development of good lubricating oils manufactured from other than Pennsylvania-grade was seriously affecting the profitability of the Pennsylvania refiners. Failure to obtain sufficient oil loomed as a long-run problem, and the owners felt that such a risk could better be assumed by a larger operator.

The agreements finally reached enabled both former owners of Freedom-Valvoline to satisfy their desires to remain interested in the oil business but to shift responsibility to another's shoulders. Bechman became a member of the board of directors of Ashland Oil. Although Craig did not become a member of the board, he remained an important adviser to the newly formed division of Ashland Oil. The informal structure of Ashland's organization permitted action to be taken in individual cases to fit the unique desires of each former owner.

THE MERGER WITH FRONTIER OIL REFINING CORPORATION

Even before the mergers with Aetna and Freedom-Valvoline were completed, Blazer was negotiating with still another company. This time it was Frontier Oil Refining Corporation in Buffalo, New York. Again the company was a closed corporation owned by two men, who had organized and managed it from its beginning to the date of the merger.

James F. Breuil and W. H. Bennett became acquainted through competitive relationships in the same sales territory while representing different major oil companies. When it became possible to purchase bulk facilities at Buffalo, the two salesmen organized Frontier Oil Works, Inc., on October 29, 1929. The company made little profit, and in 1933 the owners sold the physical assets to the Texas Company. Immediately, however, the two men purchased land in nearby Tonawanda, with cash from the sale and constructed new facilities at this location, which they still occupied on the date of the Ashland Oil merger.

The firm entered the domestic fuel oil business and through aggressive actions established the name "Frontier." Development of a burner division for domestic fuel oil was important in the growth of the firm, as service performed by the company on oil burners increased sales for domestic heating. The goodwill developed through this operation firmly established the company in the oil business in the minds of consumers of western New York.

Later the company entered industrial fuel oil marketing and in 1937 constructed a small crude oil distillation plant with a capacity of 2,000 barrels per day. Newly discovered crude oil in Michigan and Illinois provided a supply of raw material which could be handled economically on the Great Lakes. Soon after the entry into refining in 1938, the Frontier name, well known in the Buffalo area through the sale of fuel oil, became identified also with branded gasoline.

Immediately preceding World War II, the company increased refining capacity to 4,500 barrels per day and installed a thermal reforming unit. A 22-mile pipeline was built from the refinery to a connection with the National Transit crude oil line. Drilling operations in Illinois enabled the company to integrate backward into production.

In 1945 the name was changed to Frontier Oil Refining Corporation, and two years later the refining capacity was increased to 8,000 barrels per day. In 1949 the company was divided into two corporations: one a producing company under the name of Breuil Petroleum Company, and the other under the name of Frontier Oil Refining Corporation. The former company was not involved in the merger with Ashland Oil.

The thought of possible merger arose as a result of similar conditions which had prevailed in the cases of Aetna and Freedom-Valvoline: there had been pressure on the refining business in 1949; need for capital expenditures was clear; the Frontier owners saw a good opportunity to expand but did not see how a closed corporation could raise sufficient funds. Moreover, one of the owners had been in poor health and desired to lighten his responsibilities.

The managements had known each other for many years. At the time of the Allied-Ashland Oil merger the thought of a possible consolidation first developed. In preliminary meetings in 1949 the desirability of a merger was recognized. Blazer visited Buffalo in January, 1949, and looked over the refinery. At this stage the bargaining ranged between $3,000,000 and $4,500,000. The actual transaction was at a figure even lower than the original offering price.

In informal discussions a proposal of a statutory merger was the most attractive. The tax advantage of the stock transfer, the minimum amount of cash needed by Ashland Oil in the purchase, and the ability of the owners to sell a portion of the Ashland Oil stock made such a consolidation desirable for all parties concerned.

From Ashland's viewpoint, the location of Frontier's refinery on the Great Lakes fitted the transportation facilities acquired with the Allied merger. Allied tankers could move fuel oil into the Buffalo area on a favorable basis. Ashland's marketing territory would continue to be contiguous, still relatively compact, and on the "economic ridge."

Refining and marketing operations were acquired with no increase in net production of crude oil, for although the crude oil properties of Frontier were owned by the same parties, they were held by a different corporation not included in the merger.

ACQUISITION OF OTHER PROPERTIES

The Frontier merger—the last of the four major consolidations—became effective on April 1, 1950, but the expansion of Ashland Oil did not stop. The next acquisition was explained in the annual report for 1950:

In April of 1950, your Company bought a substantial inventory of petroleum products owned by the operator of the old National Refining Company plant at Findlay, Ohio, on a basis which permitted this plant to be acquired at a nominal cost. This plant with a rated capacity of approximately 10,000 barrels of crude oil per day and facilities for manufacturing and packaging automotive and industrial lubricants, asphalts, and other products, had been shut down prior to the time it was acquired. After some alterations, it was reopened on a profitable basis. . . . In addition to the refinery, the sales organization and the brands of the old National Refining Company were acquired with this purchase, including well-known "White Rose" gasoline and "Enarco" motor oils and lubricants.

This purchase is another case in which Blazer was able to purchase a large amount of obsolete equipment along with valuable goodwill and business connections at practically no cost to the company. National had been declining for twenty years, and the refinery had lost money.

National Refining Company dated back to the nineteenth century. It had been one of the leading independents that had pressed the fight against the original Standard Oil interests. The company had operated refineries at Cleveland, Marietta, and Findlay, Ohio, and Coffeyville, Kansas. Early in the present century it had acquired a Canadian interest that became very valuable and was ultimately sold in 1936.

Thomas A. Combs

J. Fred Miles
sitting on a field
storage tank

TOP: THE ORIGINAL LEACH REFINERY IN 1924
BOTTOM: POSTWAR AERIAL VIEW OF THE NO. 2 REFINERY AT CATLETTSBURG

In the 1930's the company began to deteriorate. An outside group finally gained control, ousted the president, and offered the position to Blazer. Upon Blazer's refusal, Paul Ryan became president and attempted to build the company through an aggressive promotional campaign. Ryan, after a serious airplane crash, resigned, and the period that followed became known in the company as "the march of the presidents." On one day the company had three. One of these executives began to liquidate the various assets.

In 1946 Midwest Refineries bought the wholesale portion of the business and the brand names. The refinery and related departments were rented. Later, after the refinery suffered a serious fire, the company began to fail to cover operating expenses and closed down.

The condition of the company in 1950 was explained by Blazer to the board in a letter:

National has large inventories carried on their books at approximately $600,000 under current market values. We are to take over the inventories at the prices carried on their books and are to pay $600,000 for the refinery. Theoretically, we will make enough off of the inventories to get the refinery and goodwill for our efforts. . . . We got into this situation through the fact that we have been supplying their crude oil requirements. Their contract had a few months to run, but they came to us recently asking for a cancellation advising that they were losing so much in the refinery that they would have to close down. . . . Although we have said that we attach no importance to the refinery, except for its value as scrap, I suspect that occasionally there might be a market on which we could operate profitably.

During the next four years the refinery actually remained on stream, except for maintenance, and operated profitably. After the insecurity of ten years and with the assurance of continued operations, the National organization was able to increase its business by 300 percent. Its standing in the industrial and railroad lubrication field remained high. The location at Findlay took on added advantages after Blazer was able to obtain space in several pipelines linking the refinery to Detroit. The primary reason that Findlay did not remain one of Ashland's most important refineries was due to its nearness to the Canton plant.

A second small refinery was purchased late in 1950. This was another plant which was obsolete and unwanted by other companies. In fact, word of the possibility of purchase came from an executive of a competitor who could not interest his management in its purchase. This refinery was located at St. Elmo, Illinois, near the Illinois oilfields. Its chief attractive characteristic was its storage facilities, which would be valuable even if the refinery were closed. It was close enough to St. Louis to provide terminal facilities in that area. After the acquisition Ashland Oil kept the small refinery in operating condition for several years and ran it intermittently whenever the market for its products appeared to be favorable. It was dismantled in 1953, but continued to be used for products distribution.

Blazer's negotiations for new properties came to an unforeseen end in 1950. Just after he had completed the last of the transactions mentioned above, he suffered a very serious heart attack while in California. He remained inactive for more than six months, although he resumed his letters to the board after four months. During this period the major department heads began to write lengthy letters to the board to substitute for the letters from the chairman. After Blazer began again to write his letters, he continued to enclose the letters from the major heads.

THE EFFECT OF THE MERGERS

All of the major acquisitions in less than two years formed a pattern with common characteristics. Each was interesting to Ashland Oil because of its marketing strength. Each brought an additional refinery into the Ashland Oil system as a by-product. Each was owned primarily by two parties who had a major investment in the venture. All were affected by tax considerations. All had operations complementary to Ashland's. All owners had had numerous contacts with Blazer over a long period of time through sales operations and industry meetings. Three of the four companies were of approximately the same age. None had any securities listed on an organized exchange. In each merger the tax-free stock exchange of securities was the mechanism through which the previous owners relinquished control and obtained stock interest in Ashland Oil. In each case the company's status in national financial circles made it possible for the former owners to obtain cash through marketability of Ashland Oil stock even though little cash was involved in the immediate merger transaction.

Blazer in his annual report to stockholders in 1950 summarized the place of these additions in the Ashland Oil system:

All of the foregoing additions to the properties of your Company are directly related to serving a compact and concentrated industrial and domestic market for petroleum products bounded by the Appalachian Mountains on the east, the Great Lakes on the north, the Mississippi River on the west and Central Tennessee on the south. A network of common carrier pipe lines and the company's own lake tankers and river transportation equipment tie these refining and marketing properties together to provide the flexibility necessary to meet constantly shifting demands for different kinds, quantities, and qualities of petroleum products.

In a sense, Blazer was opportunistic in his acquisitions during this two-year period of rapid growth through mergers. Yet the common characteristics did form a pattern. Furthermore, the marketing differences among these companies made them fit together in a complementary manner: Allied had an organization for heavy fuel oil; Frontier specialized in domestic fuel oils; Freedom-Valvoline and National brought a strong organization in lubricating oils; and Aetna provided additional channels for gasoline.

The rapid growth presented a great challenge to the Ashland

Oil management. Blazer in his speech to the Newcomen Society in 1956 looked back and made the following observation concerning the problems to management.

Mergers must not result in stepchildren . . . mergers frequently require the reconciling of different philosophies of business. Planning and patience are necessary in order to avoid inequities to individuals and to protect personal pride. It is not easy to bring two groups of people, who have been trained differently, into a smoothly functioning team under a consolidated leadership. The acceptance of a certain amount of entrenched inefficiency may be necessary; likewise new talent must be recognized and rewarded. A period of months may be required in which to relocate people into positions where their abilities may be better utilized and their shortcomings minimized.

6

PIPELINES AND TOWBOATS

The Transportation of Oil and Its Products

THE SERIES OF mergers in 1948-1950 not only quadrupled the value of Ashland's assets, but broadened its marketing territory from a circle extending roughly 200 miles from the city of Ashland to include all of the Ohio Valley and a substantial part of the Great Lakes area. This geographical expansion brought with it an increased problem in transporting crude oil and finished products.

The Ohio River remained the key to Ashland's transportation system, and towboats with oil barges now extended their trips upstream to Freedom and downstream to the Mississippi, which they plied from Louisiana to Illinois. On the Great Lakes, Ashland Oil became the owner of two tanker fleets, which served to connect the new acquisitions at Toledo, Cleveland, and Buffalo. No convenient north-south waterway, however, existed to join the upper tier of plants with the lower.

THE ECONOMICS OF OIL TRANSPORTATION

In the United States the major production of crude oil is located at some distance from the regions of greatest consumption.

A company's location of refineries and marketing activities is dependent upon the means available to provide the refined products to an area.

Transportation in the industry has two aspects: movement of crude oil from the fields to the refinery, and distribution of refined products from the refinery to the consumer. Both are affected by the nature of the product. Frankel stresses this unique nature: "The main consequence of the liquid state of petroleum is that it requires specialized equipment . . . specially designed for a liquid and is thus no good for anything else. The consequences are (1) the oil industry has always had to consider transport as being a major problem to be solved within its own orbit . . . (2) it is a constituent factor which has considerably influenced the structure of the industry. As a matter of fact, the development of oil economics can best be described in terms of transport."[1]

The principal methods of transportation include railroad tank cars, pipelines, ocean and lake tankers, barges, and tank trucks and wagons. Development of efficient pipelines and waterway carriers has made long hauls by railroad economically prohibitive as a primary means of transportation. In the early development of the industry the tank car was the basis for negotiations through the operation of a modified multiple-basing-point system (for example, Group Three, Gulf Coast); however, in the last twenty years the pipeline has been the primary overland reference method.

The economics of pipeline operations has had a strong impact upon the managerial decisions of oil company executives. Products move through pipelines in one direction and require no returned empties; their viscosity must be low enough to permit easy flow; a large quantity must be available at a fixed location from which the product can be pumped to a fixed destination. The unit cost of operating the line is especially responsive to the size of the line and the percentage of capacity at which it

[1] Frankel, *Essentials of Petroleum*, 33-34.

can operate. Most oil pipelines have been owned by oil companies themselves, since for an independent firm the risk of supply and distribution has been great. The nature of the operation requires that transportation be very closely coordinated with producing, refining, and marketing. Because economical use of the line is a function of the number of barrels moved, the value of the pipeline decreases rapidly when the importance of oilfields declines or markets shift. The scrap value of the pipe usually approximates the cost of salvage and makes shifting of pipeline operations undesirable.

Construction of pipelines requires great financial resources. As the scale of the line increases, the unit cost of additional capacity decreases. McLean and Haigh show that a 10-inch line will carry 45,000 barrels of oil per day at a rate of 37 cents per barrel for a 1,000-mile haul; a 30-inch line will carry 350,000 barrels of oil per day for the 1,000-mile haul at a rate of 10 cents per barrel.[2] This economy of scale is a major obstacle to the private use of pipelines by smaller companies.

In its early history Ashland Oil was not at a disadvantage as a result of these large-scale economies, primarily because it was filling a niche in a local and protected market near a crude oil supply. As the company expanded its marketing area, it began to experience competition from large operators which had important advantages. In such expansion it had five possible choices pertaining to transportation: to cooperate with others and to build and operate jointly a "large-inch" pipeline, to expand the total volume within a restricted market area in order to be able to operate efficiently a pipeline of its own, to find some other means of transportation that would enable it to enter a market on a competitive basis with oil transported by pipeline, to use common carrier pipelines, or to specialize in refined products not adaptable to mass transportation by pipelines.

Ashland's management used several of these possibilities, but

2 McLean and Haigh, *The Growth of Integrated Oil Companies,* 186.

119

most important, the Ohio River gave the company the opportunity to operate a barge fleet. There have been many estimates as to the relative costs of the various means of transportation. The answer depends upon a number of factors; however, an example offered by McLean and Haigh will give a general picture of the condition in January, 1952: from the Gulf Coast to Bayway, New Jersey, by railroad—$2.83 per barrel; by pipeline—$0.78 per barrel; by ocean tanker—$0.655.[3] The pipeline and ocean tanker rates declined significantly during the 1950's. Modern barge transportation on the Ohio and Mississippi rivers would compare approximately with the costs by ocean tanker and pipeline.

In a speech in 1949 Howard Marshall, president of Ashland Oil, emphasized the strategic importance of barge transportation to a small company such as his:

Without the rivers, we see two great oil producing and refining areas —the Mid-Continent and the Gulf Coast, each widely separated from the big consuming markets of the East and Middle West. With no inland waterways as a competitive transportation factor, only railroads and pipelines available, the rails cannot compete, so far as costs are concerned. But the building of a pipeline is a major financial undertaking and the line must be operated continuously at high volume in order to remain economical. Only the largest shippers can qualify, since only they command both the large capital resources and large volume required.

. . . . .

As contrasted with this large expenditure, it is possible here in Cincinnati for an independent marketer of gasoline to buy gasoline at Baton Rouge, Louisiana, and have as little as a few barges a year delivered to him by commercial barge line—all at a transportation cost of approximately that of his largest competitor.

Ashland's strategic use of barges after 1938 offered certain advantages over pipelines: the relatively high initial cost of pipelines could be avoided through the use of barges and towboats; barges could be shifted from one oilfield to another, dependent

[3] McLean and Haigh, 184.

only on the course of the river, whereas the pipeline demanded a fixed flow for a long period of time—this flexibility allowed the company to purchase crude oil in those areas not serviced by pipelines and to sell small quantities in areas which might temporarily yield higher netbacks; and the motive force in barge operations was separate from the cargo-carrying space and made it possible to move both crude and finished products with the same equipment by merely shifting from "dirty" to "clean" barges and vice versa.

Disadvantages of barge transportation tend to balance these advantages. As larger pipelines have entered the area, the cost advantage of the barges has become neutralized. Increased efficiency of barge transportation has depended upon governmental action in the construction of high-level dams and larger locks on the inland waterways. The use of barge equipment has required a constant effort on the part of the company to eliminate lost time waiting at locks. The barge system has been adversely affected by weather factors. Ice on rivers has halted traffic; low rainfall has made the maintenance of channels difficult; too much rainfall has caused the flooding of terminals and made clearances under bridges impossible. Coordination of inventories and scheduling of shipments, even under the best conditions, present a major problem.

It has been explained that the decision to locate the refinery at Leach, Kentucky, in 1924 was based upon its transportation advantage. This advantage resulted from the following characteristics of the location (probably in order of their importance at that time): the Cumberland pipeline, which operated as a common carrier, could deliver eastern Kentucky crude oil to the refinery at reasonable costs and in adequate quantity for the size of operations then considered; refined products could be shipped over two railroads; and the location on the Big Sandy River enabled the management to use a small barge to deliver refined products to customers in the immediate vicinity.

Traffic functions were among the first service activities in the

Ashland Oil organization. In 1931, upon the death of the traffic manager, Stokely Rion, legal and claims aspects of his functions were separated from line supervision of company-owned transportation facilities. This reorganization resulted primarily from Blazer's idea of tailoring responsibility to the capacities of available men. Such actions have had a lasting effect on the company's organization. In this instance, the transportation functions were never again centralized. Railroad traffic, river transportation, and pipelines thereafter were supervised by different superintendents, each of whom reported to the chief executive.

EXPERIMENTATION WITH RIVER TOWBOATS AND BARGES

Great Eastern Refining Company had used barge delivery on a limited scale from the Leach refinery. In his first year of operation of this plant, Blazer reported to Swiss the expansion of this means of transportation:

Our river transportation has shown a profit of $6,158.13 and we expect to further strengthen this department as it offers an excellent opportunity for profit in addition to extending our sales market. Our predecessors had only six points for water delivery of gasoline and kerosene—Today, we have twenty-one points of delivery and at Ashland we serve seven different customers through one river storage tank. Twelve of our customers are unable to receive delivery except by water and are entirely dependent upon us for their supply.

The river transportation equipment initially consisted of a small boat named the *Colonel.* It pushed one barge with five tanks mounted in the hull. Having low horsepower, it operated no farther on the Ohio River than Ironton, Ohio, and Huntington, West Virginia. The company added the *Scout* in 1926 and the *Ruth Ann* in 1930. The latter boat made river operations possible as far as Cincinnati and Charleston, West Virginia.

With greatly increased business in the Cincinnati area, Blazer

122

decided in 1935 to construct a more powerful towboat. Until that time a large percentage of the gasoline sold in the Ohio city had been supplied through exchanges with companies owning refineries in that vicinity. Blazer explained to the board in October, 1935:

It is largely a matter of having the tonnage to keep a large unit busy, but each time we perfect an arrangement for exchange of gasoline with some major oil company that operates in this territory, we are placed in a position to transport gasoline for their account from our refinery to their storage plants which very often in this area are located on the Ohio River and its tributaries. Also, the having of facilities for economical transportation from our refinery makes it easier for us to work out exchange arrangements.

The last sentence represents reasoning that often appeared in Blazer's letters. The idea of maintaining bargaining power with other companies was always considered.

In 1936 Ashland Oil completed the *Senator Combs*, the first towboat constructed specifically for its use. With its 300 horsepower the *Senator Combs* was one of the finest towboats on the inland waterways. In 1938 it pushed its first cargo-type barges (in place of hulls with deck tanks) with 380,000 gallons capacity.

The successful experiment with the *Senator Combs* prompted Blazer to look ahead to more rapid expansion in July, 1938, when he advised the board:

When we bought our 300 horsepower boat about two years ago, we were counting on down-stream towing. Because we now have loads in both directions and almost three times as much tonnage as expected, we find we need a boat having 750-1000 horsepower. We have no plans for purchasing such a boat in the immediate future, but it is something that should receive consideration as soon as we have the money available.

Until 1938 the Ohio River had served only as an auxiliary means of low-cost transportation and as a basis for possible future expansion. With the success of the *Senator Combs* and the discovery in 1937 of new fields of crude oil in Illinois, the river took on a role of major importance in the operations of Ashland Oil. Until that time, most of the crude oil supply for the refinery had come through the pipeline from eastern Kentucky. Starting in 1935, crude oil came from western Kentucky, but not in important quantities. In early 1939 the management looked to the Illinois basin for additional supplies.

Blazer negotiated with Illinois Pipe Line Company for a system which carried crude oil from wells in western Kentucky to the river at Owensboro. In preliminary negotiations, the Standard Oil Company (Ohio) showed interest in a joint purchase. Blazer decided to act independently when the Standard management did not make a decision promptly. He explained to the board the reason for this quick action:

I had planned to make a proposal to the Standard Oil Company to combine the operations of their line with the one we are purchasing, but I recognized that it would put us in a much better trading position to own the line and then negotiate with them than to have to negotiate with them in advance, since if we should fail to come to an agreement as to the value of the sohio line, they could leave us out entirely. Obviously, Illinois Pipe Line Company, a part of the original Standard Oil group, would give preference to Standard Oil Company (Ohio), as compared with us.

In March, 1939, Owensboro-Ashland Company began to operate the newly purchased pipeline. Later, 50 percent of the stock of this company was acquired by the Standard Oil Company (Ohio), and their Fordville system was placed in the jointly owned company. Blazer indicated the relation of this purchase to the increase in river equipment: "The acquisition

124

of the Owensboro Pipe Line argues in favor of our reviving our interest in the early purchase of a larger towboat."

Increased use of the Ohio River during 1939 made it advisable to purchase a number of new barges to replace those which had been chartered from other companies. The *Senator Combs* was proving a valuable asset, but by this time it was towing only approximately 40 percent of the total tonnage transported on the river by the company; the remainder was handled on a contract basis.

Blazer's decision to place large sums into capital equipment for river transportation had a fundamental effect on future growth. Other companies were using the river at an increasing rate, but many were leasing rather than buying their equipment. Allied Oil Company and Aetna Oil Company had used the river, but neither had developed an integrated river transportation operation.

Its success with the *Senator Combs* had convinced the Ashland Oil management that it could operate its own fleet, tailored to its needs, at a much lower cost than if the work were done by contract carriers at the average rates. In August, 1939, Blazer summarized the position of the company to the board of directors:

Steadily over the past few years, we have been losing some of our geographical advantage. Pipe lines have come into our territory both from the East and from the West and the advantages of water transportation are being used more extensively by the larger companies to get their products into this immediate area. At times I can look out of my office window and see barges moving down the Ohio to Cincinnati carrying gasoline which originated at refineries on the Atlantic Seaboard and also see barges moving up the river carrying gasoline from points on the Mississippi River as far south as Baton Rouge, Louisiana. In fact, some gasoline from Gulf Coast points moves as far north as Pittsburgh.

Of even more competitive significance is the increased number of exchanges of gasoline by major oil companies. Many companies which shipped into this territory from far distant refineries now receive their supplies from the Pure Oil Company plant at Charles-

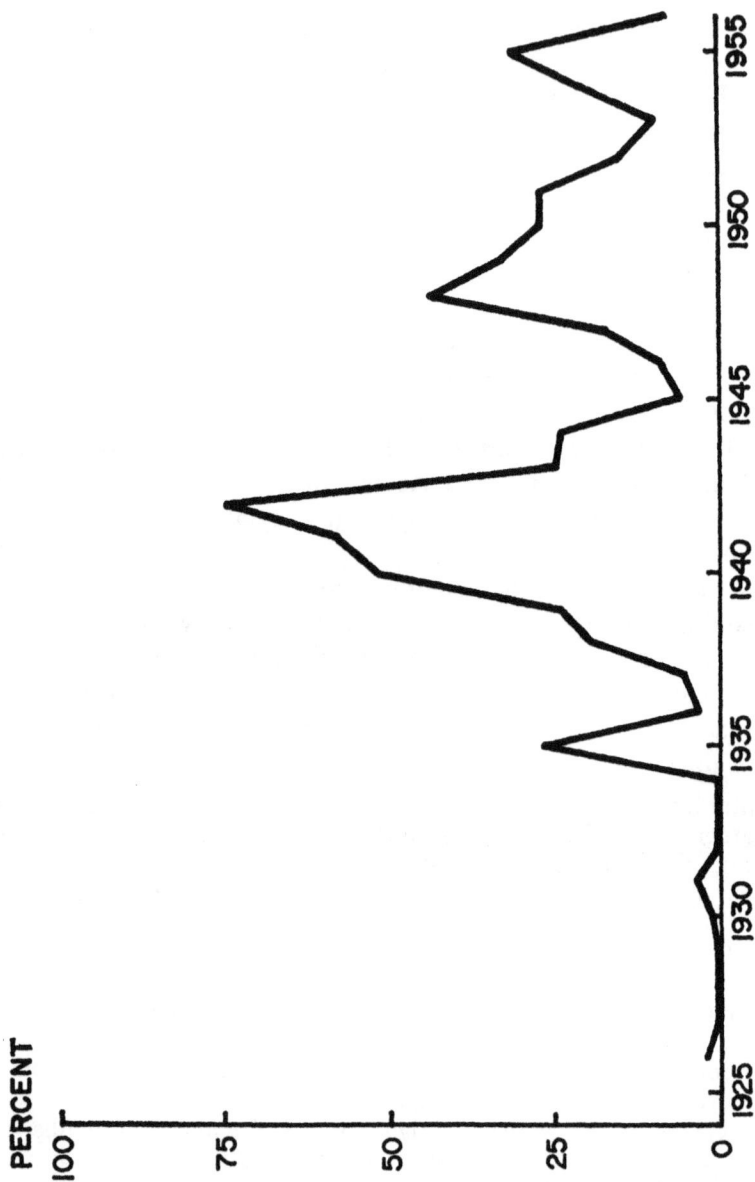

EXHIBIT 16—PERCENTAGE OF TOTAL CAPITAL EXPENDITURES PLACED INTO TRANSPORTATION, ASHLAND OIL & REFINING COMPANY, 1924-1956*

* Purchase of Cumberland pipeline not included in 1931.

ton, West Virginia, or the Standard Oil at Latonia [Kentucky] or Gulf Refining Company in the Cincinnati district.

. . . . . .

We are continuing to show larger earnings on our investment than our larger competitors, but I am not sure that we can continue to do so. The time may not be far distant when we are down to a return of 5% instead of our earnings for recent years which averaged between 10% and 15%. As you probably know, most of the major oil companies, over the past few years, have earned between 3% and 5% per year. I can perceive many competitive conditions which will tend to encroach further on the favorable situation which we have been able to enjoy for a number of years past. In fact, already we have been forced to give up some of the advantages on which we have capitalized, but usually we have found some other way to compensate for what we have lost; this past year it has been our increased volume of business. I do not feel that we can continue, indefinitely, to do that. On the other hand, I am not in the least discouraged concerning our prospects, as I see no reason why we should not be able to continue to do as well as our competitors.

The first more powerful diesel towboat added to the fleet was the *Jim Martin,* in May, 1940. Three additional boats of still greater horsepower (1,600) went into operation in rapid order: the *Ashland* in May, 1941, the *Paul Blazer* in December, 1941, and the *Tri-State* in April 1942. Each of these three boats cost approximately $300,000, not including the necessary new barges. The decision to add the *Tri-State* was based upon possible future need rather than upon present requirements. Blazer commented to the board in 1941:

Although I cannot see the definite need for such an investment, my judgment tells me that we should build still another towboat and barges to accompany same. We have a verbal option, good for another week, on another boat [*Tri-State*] at the same price as the one now under construction. . . . River equipment, such as we are buying, could be leased out today on a rental basis which would pay out in thirty months.

Thus it will be seen that immediately preceding World War II, Ashland Oil increased its capital expenditures in transportation. Exhibit 16 graphically shows Blazer's idea of quickly adjusting to new circumstances. During this period he made maximum use of his authority to decide issues on the spot that might easily have become entangled in channels of correspondence and committee meetings in a more formally organized structure.

Additional terminals were included in these expenditures on transportation. Terminals served important functions for a system of the type that Ashland Oil operated. Uncertainty of arrival of boats because of bad weather or poor river conditions necessitated having terminals sufficiently large to handle reserve stocks of crude and products to maintain a continuous flow to the other operations. Poor coordination at this critical point would have caused serious problems for the entire company. Terminals also operated as a decentralized storage section of the refinery and served in lieu of bulk stations as a means of decreasing the marketing costs of refined products. Further, they provided the link at which the means of transportation was changed. Delivery by river barge and tank truck required terminals at the point of transshipment.

During the first six years, the company's only terminal was located at Leach. Later, Tri-State provided terminal facilities on the Ohio River even after the refinery was dismantled. In several other instances, purchases of refineries were made primarily in order to acquire favorably located terminals.

In 1939 the operation of all river terminals was placed under the transportation department. Although the expanded use of such facilities constituted functions closely similar to refining and marketing functions, the coordination of this nerve center remained in transportation. Until a new functional department was organized in 1954, the transportation department coordinated production planning and control for the entire company.

The importance of the new investment in transportation was stressed by Blazer in a number of letters to the board of directors. A sample, written in 1940, follows:

THE AETNA-LOUISVILLE
HEADING DOWN THE OHIO
RIVER BELOW ASHLAND

LAYING A PIPELINE
ON A STEEP HILLSIDE

TOP: A PEPPER SERVICE STATION IN THE 1930's
BOTTOM: AN ASHLAND SERVICE STATION IN THE 1950's

I anticipate that this boat [*Jim Martin*] which represents less than 4 per cent of our total investment, will account for approximately 20 per cent of our company's operating profits for the coming year. . . . If we avoid being in the red, it will be only because our transportation profits are large enough to offset our refining losses.

In addition to these offensive steps through the construction of new towboats and barges, the Ashland management used defensive measures for its river equipment through its opposition to refined products pipelines. The federal government considered the construction of large-diameter crude and products lines running from the Texas coast to the eastern seaboard. During this period Blazer was a leader in opposing the route for these lines through the Ohio Valley, favoring, instead, the shorter route through Georgia and the Carolinas. The final decision of the Petroleum Administration was in favor of the northern route.

Later during the war, the "Big Inch" and "Little Inch" pipelines caused the Ashland management to consider their potential effects on the competitive picture after the war. In 1943 Blazer commented to the board:

In addition to the 24 inch pipe line now being built from Texas to the East Coast which we consider a serious threat to the profitability of our postwar operation, it now appears that the same group will construct for the account of the Defense Plant Corporation a 20 inch line from the vicinity of Houston on the Gulf Coast to the Eastern Seaboard. There is considerable controversy between the major companies who are sponsoring the line as to whether it will parallel the first or come up through Tennessee, Kentucky, and West Virginia passing within five miles of our plant. . . . The seriousness of this threat will be realized when it is explained that the new line will transport gasoline from Texas into this area at a cost of not exceeding ¼ cents a gallon whereas the present prevailing price at Texas refineries is approximately 1½ cents a gallon lower than the price f.o.b. our refinery. It is easily conceivable that this pipe line built at government expense may completely destroy the refining industry in this general area after the war. Texas has large reserves of crude oil which can be produced at lower cost than can be produced in any other area. I have no solution for the problem confronting us which you can be assured will be the most important factor determining our plans for the future.

The importance of these actions in the transportation branch cannot be overemphasized. The river equipment available to Ashland Oil during the wartime period permitted the company to continue to obtain crude supplies at a time when other independent oil companies were operating at less than capacity for lack of sufficient crude oil. The acquisition of large towboats and barges provided critically needed transportation. It can be assumed that if the crude oil supply had not been clearly available, the government later would not have constructed its large high-octane plant at Leach; without the Leach "refinery No. 2," the later expansion by merger would have been less probable.

EXTENSION OF TRANSPORTATION THROUGH MERGERS

No new river towboats were added immediately after the war; however, the existing boats were supercharged and modified to carry much greater payloads. Additional terminals to strengthen the company's marketing position were the primary postwar transportation efforts.

In April, 1948, Ashland Oil purchased controlling (one-third) interest in Southern Pipe Line Company at a cost of $150,000. Although this 200 miles of trunk pipeline which extended west through Pennsylvania from the eastern seaboard was not very useful to the company at the time, Blazer decided to buy it, he explained in a letter, because of the potential use of the line at some future date in the event that foreign crude became available at economical prices for use by the Ashland Oil refineries. In further preparation for the future, Ashland Oil purchased jointly with Allied Oil Company two oceangoing tankships to be operated by a separate corporation, Allied-Ashland Tankers,

Inc. This joint purchase took place during the middle stages of the merger negotiations between Allied and Ashland Oil.

The merger with Allied expanded Ashland's transportation system significantly. Allied's subsidiary, Central Pipe Line Company, which operated 170 miles of gathering lines principally in Illinois, increased the security of Ashland's crude oil supply. Allied also owned two tanker fleets on the Great Lakes. Cleveland Tankers, Inc., had four ships, and Lakeland Tankers Ltd. (Toronto) two. These six tankers operated for Allied's account between company-owned products terminals at Cleveland and Erie and a crude oil terminal at Toledo, and between refineries and terminals of others. Allied had built a heavy fuel oil river terminal at Follansbee, West Virginia, and this, too, fitted into Ashland's existing river fleet.

In the Aetna merger, Ashland Oil acquired 93 miles of gathering lines in the Illinois basin, 33 railroad tank cars, and a number of tank trucks. With the Freedom-Valvoline merger, the company secured 1,400 miles of pipelines for gathering Pennsylvania-grade crude oil. With the Frontier merger, it obtained a large fleet of tank trucks and a crude gathering system in the Illinois oilfields.

In all of the mergers, Ashland's transportation system formed the flexible bond around which the other companies were developed into a well-balanced, unified operation. The transportation system enabled the sales organization to expand the market territory to include most of the area shown in Exhibit 5.

By July 31, 1951, Ashland's transportation system had grown to such an extent that the net investments in transportation facilities were greater than the net investments in any one of the other three branches. Net fixed assets, July 31, 1951, as obtained from the prospectus of the company dated October 1, 1951, showed

Production $13,309,074
Transportation 17,881,838
Refining 11,848,938

131

Marketing 5,375,716
Miscellaneous 3,179,680
$51,595,246

The company's total assets in transportation included all facilities with the exception of Owensboro-Ashland Company, which appeared only as the net amount invested in the company. Any comparisons with larger companies as to the percentage of transportation assets owned would be misleading, since many pipeline companies are jointly and equally held by two oil companies and are not consolidated on the books of either owner.

PERIOD OF REEMPHASIS ON EXPLORATION AND DEVELOPMENT

The Korean War in 1950 changed the outlook for crude oil and again introduced the factor of the excess profits tax. Blazer decided once more to expand Ashland's exploration department and to make adjustments in its organization to meet the new situation. This time a new oilfield in west Texas appeared to promise a real opportunity for the company. The annual report of 1951 explained:

Of primary importance in the further expansion of the company into the producing branch of the oil industry has been the acquisition of slightly less than one-half of the operating interest in leases covering some 19,400 semi-proven gross acres in the Spraberry area in West Texas. These large undeveloped holdings are believed to provide the Company, for the first time, a favorable opportunity to invest over a period of years, a substantial part of its gross profits in the development of crude oil reserves.

This Spraberry project was a major effort on the part of Ashland Oil to integrate into production. It absorbed approximately $15,000,000 over the years 1951 and 1952. Difficulties in bringing the expected volume of oil to the surface, however, necessitated a writedown of assets by $8,500,000 at the end of

132

1952 and reduced retained earnings by which expansion could be maintained. This was the first of several factors which ushered in a period of retrenchment in 1953.

Interest in the Spraberry area encouraged Blazer to take part in the construction of the Tex Harvey pipeline system to transport oil economically from the producing wells to common carrier lines. This pipeline, too, turned out to be a financial headache because of the rapid decline in production from Spraberry wells.

The company extended its exploratory operations into the Denver basin and other areas in the West through cooperative efforts with producing partners. In November, 1951, Murphy Corporation of El Dorado, Arkansas, and Ashland Oil joined in the formation of an exploration company, Amurex Oil Development Company, which acquired extensive leases in western Canada and provided Ashland Oil with a new means of increasing its interest in producing properties. Ashland Oil held 40 percent of the class-B common stock with an option to convert to class-A stock in the event of important oil discoveries. In this way the company made use of the executive talent of a producing specialist, secured additional funds from the public, and continued to expand its exploratory operations. Amurex and American Independent represented the two early instances in which Ashland Oil looked to foreign oil in its expansion. Amurex failed to discover much oil, but American Independent was successful.

During these years Blazer's policy was to purchase crude oil in excess of current requirements. Often Ashland's efficient barge system earned a good profit by moving crude oil from areas of large supply to areas in which the oil was more valuable. The ever-present threat of inability to secure a sufficient supply of crude oil made it desirable to maintain excess supplies. As a result, the company acted as a seller of crude oil on many occasions.

By the end of 1952 the production department had built an organizational structure for a long-range exploration program.

Disappointments of major proportions, however, caused a change once again in the manner in which these plans were actually carried out.

PERIOD OF DEEMPHASIS ON EXPLORATION AND DEVELOPMENT

During January, 1953, reports of the recovery of crude oil in the Spraberry area caused a fundamental change in the actions of the management. The annual report of 1953 explained the 1952 effort to place exploration on a long-run basis:

Until recently, the company's crude oil producing operations had been conducted on a somewhat opportunist basis with particular regard for tax benefits. During 1953 the company reorganized and strengthened its crude oil exploration and production department and gave greater emphasis to a long-range program.

However, after this long-term program was set up, Blazer reported to the directors in December, 1953:

I am concerned about the continued losses in our producing department. These more than offset the profits that are accruing currently as a result of improved showing in the Spraberry area and substantial water-flooding in Illinois. I am afraid that we can't afford to continue to subsidize our producing operations, since it appears that refinery profits are going to be smaller. I am sure that we are better organized in our producing department and that Roy Ralston [production manager] has better control of what is going on, but that doesn't help much if our production continues to decline in the face of increased expenditures. The current accumulation of wildcat acreage and geological data may pay off eventually, but we can't afford to carry our division offices that long at a loss. We may find we have to spend our money where we can obtain returns more quickly.

Crude oil became so plentiful at this time that the Texas Railroad Commission placed additional restrictions on its production. Ashland Oil had a crude supply sufficient for the short run; in fact, the large supply had become a problem, as Blazer explained in a July, 1954, letter:

We have a crude oil supply that is so large that it is burdensome. It is almost as difficult to cut off crude oil supply as to increase it when additional supplies are needed. We buy approximately 65,000 barrels per day of oil direct from producers on lease division orders.

Again the facts illustrate that flexibility in crude oil supply was hard to maintain, even if the company purchased its major portion and owned little net production. The structure of crude oil markets makes it very difficult to obtain a supplier again once the buyer had refused to purchase the oil. The reputation of the purchaser in the oilfields was most important when crude was scarce and when that purchaser was dependent for a large share of his supply upon independent operators. A company with small net production did not control the quantity of crude oil brought to the surface and thus could not prevent an over-supply. Through greater control over production stages, the integrated firms probably possessed greater flexibility in the adjustment of crude oil supply to refinery needs than noninte-grated firms.

In spite of the large supply of crude oil above ground, Blazer experimented with water-flooding old wells in the eastern Kentucky fields. This secondary recovery program continued to be one encouraging aspect of Ashland's production of crude oil. Its objective was to increase production in these fields over a long period of time to approximately 7,500 barrels per day, the capacity of its pipeline facilities in the area.

Although new wells began to flow in Oklahoma, the western operations did not show profits. The company carried out a drilling program that had been planned previously, but gradually retrenched in these operations. In February, 1956, Blazer observed:

I am gradually arriving at the conclusion that we are not equipped for profitable exploration, except east of the Mississippi River.

Apparently, we lack the advantages of small independent operators without having the organizational advantages and experienced personnel of the larger oil companies.

In August he added:

It is difficult to get across to our organization that, even though we are in a comparatively large company in refining and transportation, we are just a little company in production and we must engage in "poor boy" operations instead of taking on the ways of major oil companies. We have less trouble in that respect in the Illinois field, where most of the other operators are small independents.

MODERNIZATION AND CONSOLIDATION OF REFINERIES

After the rapid acquisition of old refineries, Blazer realized that large sums would be required to modernize equipment. Experience from the past operation, however, enabled the company to continue running these small obsolete refineries profitably for several years.

In 1950 Ashland Oil owned only one catalytic cracking plant (at Leach). Blazer chose Canton (the former Allied refinery) as the first in his program of major capital improvements. By 1952 the management had increased throughput of the Canton refinery to 30,000 barrels per day and had completed a fluid catalytic cracking unit. Later, another "cat cracker" was built at the Buffalo refinery and went on stream in 1953.

Modernization was based upon the assumption that a refinery of approximately 25,000 barrels per day was the optimum size for Ashland Oil to operate.[4] Moreover, successive improvements in one refinery at a time both suited the financial capacity of the company and permitted it to take advantage of technological

---

[4] McLean and Haigh, 585. "One study revealed that by following certain practices . . . a small refiner could have built a modern 25,000 barrels per day refinery in 1953 and earned a percentage return on his investment approximately the same as a large company with different operating practices could have realized on a new 100,000 barrels per day plant."

improvements of the moment. The recognition of the tremendous impact of obsolescence in the industry remained important. McLean and Haigh emphasized this change when they observed: "The development of refinery processes has been highly dynamic, and the state of the refining art has been undergoing change at a seemingly ever-accelerating pace. It was a widely held view among 'refiners' in 1952 that it was impossible to build a completely modern refinery. Although all the latest developments might be embodied in the design of a new plant, the refinery would inevitably be obsolete in some respects before it was erected and on stream."[5]

By 1954 the Catlettsburg catalytic cracking unit had become relatively obsolete and in need of capital expenditures. A new catalytic reformer was built to help the company maintain the pace of the "octane race." Blazer by this time felt that the optimum size of refineries had increased, as he wrote to the board in May, 1954:

I believe that small or what heretofore have been considered medium size refineries are reaching the end of their economic life. Approximately the same number of men are required to operate a small processing unit as a large unit. . . . I presume the situation at National [Findlay, Ohio, refinery] is not unlike that of Tidewater Associated Oil Company at Bayonne where they are abandoning a complete refinery of comparatively large capacity because it is preferable to build elsewhere and operate a completely new plant than attempt to modernize the present plant with the related manpower problems.

This observation is of particular significance, for it expressed an opinion that the past techniques of employing small, old refineries would be inadvisable in the future. This fundamental change and the fact that the company had become a much larger operation demanded a number of alterations in managerial and operational techniques. In interviews in April, 1954, Blazer recognized that the situation faced by the company required modification of past methods.

[5] McLean and Haigh, 537.

The Spraberry disappointment started a period of consolidation not only in exploration but also in refining. Weakened markets and the increased demand for higher octanes brought a reappraisal of all of the company's refineries and consideration of the possibilities of closing the less efficient plants. The small St. Elmo unit had been dismantled in 1953, and the Niles plant was placed on a standby basis for intermittent operations depending on the steel industry's market for residual fuel oil.

Blazer decided to discontinue operations at the older Leach refinery, known as No. 1. Most of the improvements that had been made at Leach had been applied to the modern government-constructed plant. The fact that some of the oldest employees in length of service in the company were affected made the decision especially difficult for Blazer, who had nursed the equipment for thirty years. He explained to the employees that the refinery was inefficient in 1954 and that it had not been advisable to make additional investment in the plant because of labor difficulties which had developed. He alleged that in these years, newer employees had obtained control of union negotiations because of apparent apathy on the part of the older employees, who had always cooperated with the management.

The management continued to study operations of other older refineries at Louisville and Findlay. Both had thermal cracking but no catalytic cracking facilities. Both had geographical advantages which permitted them to continue to operate on a marginal basis. Findlay was located on a products pipeline to Toledo and Detroit; however, the location was so close to the newer Canton plant that its area of distribution was limited. Louisville was on the Ohio River and in the middle of a good marketing area; however, an air pollution campaign by the city and the uncertainty of future competition from low-cost, large-diameter pipelines discouraged any major capital expenditures. In October, 1955, the Louisville refinery was closed down and used only as a river terminal.

138

The trend toward production of higher octane gasoline and the decrease in the rate of growth of demand for petroleum products directed Blazer's attention to decreasing unit refining costs in ways other than by increasing throughput. The major capital expansion during this period, therefore, was not to augment greatly the company's throughput but to improve facilities for producing higher octane gasoline and to effect economies. Refining continued to be the branch which received greatest emphasis by Blazer. The idea of maintaining flexibility in management and operations was especially valuable in this branch. The past success of the company was in no small part the result of its efficiency in refining and the adaptability of production to consumer demand for multiple products.

The five choices pertaining to transportation, listed earlier in this chapter, remained important to the management in the 1950's. Competitors were cooperating in the operation of pipelines. The construction of the Mid-Valley pipeline, which brought crude oil into Ashland's marketing area from the Texas fields, had seriously challenged Ashland's low-cost barge transportation. Ashland Oil participated with 5 percent interest in the El Rancho pipeline, a 24-inch line from west Texas to the Gulf Coast, as a means of delivering its expected increased production in Spraberry to its water transportation system.

In 1953 Blazer felt that it was especially difficult for him to obtain cooperative undertakings with his competitors because of their dislike for Ashland's practice of selling to price marketers. Nevertheless, he continued to make efforts to arrive at agreements beneficial to the Ashland Oil transportation system. Although some companies might have desired to cooperate with Blazer, the geographical location of Ashland's refineries were such that he generally would have wanted the flow in products pipelines to be in the opposite direction from that desired by the other companies. Ashland Oil had acquired small refineries located away from refining centers and close to markets not easily supplied from these refining centers. Possibly, other firms had not wanted the refineries acquired by Ashland Oil for the very reason that they were not at primary centers. The advantages these refineries had held for a small company were slowly becoming disadvantages for large-scale operations.

By 1956 Ashland Oil had become more generally accepted in the petroleum industry as a large company. It had increased the capacity throughput of its three main refineries at Canton, Buffalo, and Catlettsburg to a total of 140,000 barrels per day. This quantity demanded that Ashland's management expand its system of crude oil transportation. Everett Wells, shortly after becoming president of Ashland Oil in 1957, explained to the New York Society of Security Analysts:

Historically, our Company has been able to acquire, build and operate economically crude oil trunk and gathering systems strategically located to enable us to deliver crude oil to our refineries at satisfactory costs. The extent of these pipelines can best be appreciated when you realize that in total mileage of these lines we rank among the top ten companies in the petroleum industry. . . . We maintain a modern river fleet, consisting of 101 barges and nine towboats ranging from 1150 to 4800 h.p. . . . This equipment brings in 85% of the crude oil requirements of our largest refinery at Catlettsburg and delivers 80 to 85% of the refined products made at Catlettsburg.

In spite of this large system of pipelines and company-owned boats, Ashland Oil direly needed to be accepted in some of the many joint projects of pipeline construction. But not until 1956 did Blazer succeed in participating in the construction of a large-inch crude pipeline which could serve its refineries.

The Tecumseh pipeline, organized by Sinclair and the Pure Oil Company, each with 40 percent ownership, and Ashland Oil, with 20 percent, was a 20-inch line with an initial capacity of 100,000 barrels per day running from Griffin, Indiana, east to Cygnet, Ohio. This type of financial arrangement is so common in the transportation programs of major oil companies that the significance in Ashland's growth might be overlooked. It was possible in the Tecumseh transaction for Ashland to acquire 20 percent use and ownership of a pipeline costing more than $15,000,000 for the total cash outlay of less than $500,000. In order to qualify for inclusion in a group which can secure this favorable type of loan, a company must have a secure source of crude oil which can be supplied to the pipeline and a stable market at the other end. In the past, Ashland's policies and scale of operations were not suitable to this type of joint effort, but in 1956 the company no longer was "a poor boy" in pipeline transportation.

USE OF THE RIVER AS A SUBSTITUTE FOR PIPELINES

Even after the company had become larger, towboats and barges remained the lifeline for most of Ashland's operations. Crude oil came from Louisiana and Illinois via the Mississippi and Ohio rivers to refineries at Louisville and Ashland. Even the refineries at Canton, Findlay, and Buffalo made indirect use of river transportation through the choice of bringing Illinois basin crude oil to those refineries through pipelines or of sending it to Ohio River refineries by water.

The greatly increased need for crude oil required additional river equipment, both to carry a larger volume and to seek supplies from more distant points. In 1954 the company purchased a towboat from the Standard Oil Company (Ohio), renaming it the *Valvoline*. During 1951-1952 it added to its river fleet two 4,800-horsepower towboats, then the most powerful on the inland waterways. These two boats, the *Aetna-Louisville* and the *Allied-Ashland*, increased the total towing capacity of the company's system by one-third. All previously owned

141

boats continued in operation with the exception of the *Senator Combs,* which the company had sold in 1949 because it had become too small for efficient operations. In 1955 two additional diesel towboats, the *Winchester* and *Cherrystone,* were acquired.

By 1956 the company's river fleet had become one of the largest on the inland waterways and was carrying 2,500,000,000 ton-miles of business. It was for this reason that Blazer exhibited concern over the proposal by the Hoover Commission to initiate user charges on the inland rivers. The commission report gave advance warning of possible legislation by Congress, and as a result, the Ashland Oil management actively supported the gathering of facts and opinions which would provide arguments against any change in the manner in which the waterways had been handled for 150 years. The Ohio Valley Improvement Association, which had been successful in promoting the building of several high-level dams on the Ohio River to increase greatly the efficiency of water transportation, concentrated opposition to user tolls.

The issue became active in 1955 and 1956. At the hearings, the railroads through their influential lobby in Washington aggressively pushed for passage of legislation. Blazer testified in opposition to the tolls, concerning which he commented in a letter to the Ashland Oil directors:

We can't afford to take a position which is antagonistic to them [the railroads], although we consider that they will not blame us for opposing river tolls which would be so costly to our Company. . . . Our position is that since we use all forms of transportation—rail, water, pipe line, truck and air—we are in favor of anything that will reduce our transportation costs and opposed to anything which will increase our costs.

USE OF COMMON CARRIER PIPELINES

The existence of common carrier pipelines can encourage operations by smaller refiners. In most cases, however, pipelines have

142

been built, operated, and used principally by the owners. The question of whether pipelines should be treated legally as common carriers or as plant facilities has been the subject of much legislation and many court decisions. George S. Wolbert has summarized the pipeline issue in his thesis: "It will be helpful . . . to recognize the fundamental inconsistency of pipe line status as plant facility by nature, common carrier by law, and a combination of the two in fact."[6]

In the light of the confusion in the status of pipelines, the management of Ashland Oil refrained from depending too greatly on the common carrier status of any pipeline. Although many believed that the problem of the small refiner became less critical when a court decision gave him the right to space in pipelines, Blazer recognized that many ways remained in which a large company could abide by the law but in effect limit the usefulness of the line to a refiner who tried to enter. Size of tender of oil, detailed specifications of quality, required terminal facilities, and other factors were potential discouraging influences to a small refiner in his attempt to depend upon common carrier status. As a result, Blazer for many years placed little emphasis on the common carrier status of pipelines.

With the newly acquired refineries at Canton, Findlay, Buffalo, and Freedom, Ashland Oil became more dependent on common carrier lines for their crude oil supply. Even in the delivery of refined products, common carriers became of greater importance after acquisition of these refineries. Common carrier lines connected the Findlay refinery with Toledo and Detroit. These common carrier lines were connected at Findlay with Ashland Oil's products pipeline which extended to Findlay from the Freedom refinery on the Ohio River, near Pittsburgh, by way of the Canton refinery. The greater part of Ashland Oil's products pipeline across Ohio consisted of an old crude oil

[6] George S. Wolbert, *American Pipe Lines: Their Industrial Structure, Economic Status, and Legal Implications* (Norman, University of Oklahoma Press, 1952), 113.

143

pipeline which had been converted to products use. New pipelines at either end had been built to connect with the Findlay and Freedom refineries. Also, the eastern end of this pipeline system was connected with a products pipeline, owned by Buckeye Pipe Line Company, which delivered Ashland Oil's products into Ashland's new products terminal in the Cleveland area. Finished and semifinished products from the refinery near Ashland, Kentucky, often moved by water to Freedom and were then pumped direct from barges into large storage tanks at Cleveland, Canton, Findlay, Columbus, Toledo, Detroit, and often as far west as Huntington, Indiana, near Fort Wayne. On occasion, products from the Ashland, Kentucky, plant might move by water from Toledo to Buffalo.

IMPORTANCE OF THE TRANSPORTATION BRANCH

The contribution of the transportation branch to the success of Ashland Oil over the thirty-three years was equal to that of the refining branch. Early operations on the "economic ridge" of transportation costs provided the company with a protected marketing area for growth. In the 1930's the Ohio River became an essential factor in the economical handling of both crude and refined products. Ashland's growth after World War II through mergers was influenced greatly by its transportation system. The future success of the company partially depended upon the manner in which the company would be able to adjust to the trend toward use of large-diameter pipelines for crude oil and to the increased use of products pipelines in its marketing territory.

144

7

JOBBERS AND BRAND NAMES

The Pressure for Markets and Promotion

THE PRINCIPAL products which are refined from petroleum—gasoline, domestic fuel oil, diesel and gas oil, asphalt,
lubricants, kerosene, and industrial fuel oil—are sold in fairly
distinct and separate markets and often through different channels. The integrated refiner uses the refinery gate as a reference
point from which he considers each product and each market
on a marginal basis. "Netbacks" on products to different locations offer him quantitative data for making his complex
decisions. These decisions are based primarily upon two factors:
the price level in each local market, and the transportation cost
for the refined products to each market.

The refiner considers three levels of markets in which he can
sell gasoline: the refinery wholesale or tank-car market, the local
wholesale or tank-wagon market, and the retail or service-
station market. Efforts to integrate forward into marketing involve increased activities in the local wholesale and retail
markets.

PRICING POLICIES

The best way to relate the pricing policies of Ashland Oil is to
quote a statement made by its chief executive:

Originally, Ashland Oil sold all of its gasoline through independent jobbers, who received deliveries by tank car or by barge in tank-car quantities. Gasoline was priced at the wholesale tank-car level to meet competition from more distant refineries. In the early years the price ordinarily was escalatored on the Group Three Oklahoma market plus a transportation charge, which usually was enough under the actual freight cost from Tulsa to discourage distant competition, but high enough to give a netback to the Leach refinery which would give Ashland Oil a better price realization than could be obtained by any more distantly located competitor. Gasoline usually was sold on annual contracts subject to automatic renewal unless canceled on sixty or ninety days' notice. When competitive conditions justified lower prices than provided by the annual contracts, Ashland Oil frequently gave voluntary price reductions. Later, when new price basing points developed at nearer refining centers, destroying much of the significance of Group Three Oklahoma prices, the company entered into contracts with hundreds of jobbers which specified that the gasoline price would be Ashland Oil's market price on date of shipment for that quantity of gasoline to that class of trade in that area. Ashland Oil was able to obtain such price clauses only because the company, over a period of years, had gained the confidence of their jobber distributors.

In order to qualify for a tank-car price a customer might not be located on a railroad track and thus could not receive gasoline by tank car, but under Ashland Oil's pricing policies he would be entitled to the tank-car price if he had a storage tank sufficient to receive the equivalent of a tank car of gasoline by barge delivery. In one instance, Ashland Oil laid pipelines from the river to a number of large-volume service stations located in the center of the business district. Each of those stations had large enough storage tanks and took deliveries in quantities equal to or larger than tank-car shipments. Gradually, over a period of years, industry practice developed of giving the tank-car price to operators of large filling stations which put in tanks large enough to hold a tank car of gasoline, even though actual delivery was by large transport trucks. For many years, Ashland Oil owned no tank trucks and made sales only to customers who could receive deliveries by either rail or barge.

Unlike many independent refiners, Ashland Oil for many years sold no gasoline through so-called tank-car resellers or brokers. By developing its own sales organization calling directly on independent jobbers, Ashland Oil, unlike most independent refiners, was able to

146

avoid selling its products through tank-car marketers, who often bought gasoline at distress prices—sometimes from major oil companies.

In the early years, Ashland Oil sold much of its residual fuel oil through tank-car marketers such as Allied Oil Company and other companies which specialized in the sale of fuel oil to the steel industry and other large users. It was not many years, however, until the company adopted the same policies in that respect as regards the sale of residual fuel oil that it had historically used in the sale of gasoline. The sale of specialty products, such as naphthas and solvents, continue to be sold to a certain extent through specialists in that field, such as American Mineral Spirits Company and The R. J. Brown Company, which, incidentally, Ashland Oil acquired in 1956.

Many refiners entered into gasoline contracts which specified that the price should be a certain margin under local gasoline tank-wagon markets. Ashland Oil avoided such contracts, since they took away from local distributors the incentive to avoid local price wars. Ashland Oil, instead, had somewhat similar contracts which specified prices based on statewide tank-wagon markets, in the case of Ohio, or on "normal tank-wagon markets" where there were published normal markets.

In later years, when Ashland Oil began distributing through local bulk plants and selling at tank-wagon prices to consumers, dealers, and service stations, including operators of company-owned stations, the pricing was on a conventional basis comparable to that of major oil companies. A dealer who desired to be unconventional in his retail prices was encouraged to market under his own brand name.

After the first few years, Ashland Oil gradually began to sell an increasing quantity of products to so-called "price marketers" who usually owned a number of service stations and had facilities to receive deliveries in tank-car quantities. Even though these marketers usually sold their gasoline for less than the prevailing prices for branded gasoline of the same quality, Ashland Oil did not give them a lower wholesale price. In fact, the company was proud of its uniform pricing policy which provided for the same price on the same quality of gasoline to all wholesale buyers in a given area, including its own local bulk plant.

Ashland Oil has always thought that more independent refiners failed because of bad pricing and marketing policies than due to any other disability.

In 1928 Everett F. Wells, who had been hired by Ashland Refining Company directly from college in 1926 and who in 1957 became president, was made assistant sales manager (Blazer was the sales manager). Yet numerous comments in his letters and speeches indicate that Blazer was finding through experience that integration was being forced upon him by competition. As early as 1927, he observed in a report to the board:

Much as I personally dislike the tediousness of retail distribution, I think that it is essential to the continued success of our company. . . . So far as I know, our company is the only refiner of Kentucky crude that ever paid a dividend out of refining profits. We are the only refiner of Kentucky crude not having marketing facilities that has been able to continue operating. With the exception of some small Pennsylvania refineries specializing in the manufacturing of lubricants, we are, I believe, the only independent refinery east of the Mississippi not having a retail or tank wagon marketing division. More and more the competition is for retail outlets. Some of our best customers have been absorbed by other refineries. . . . Expansion into marketing is a form of insurance to protect our present very profitable investment.

Because of the pressure to integrate forward, most of the numerous purchases in this period, summarized in Exhibit 1, were of local wholesalers. In January, 1929, Blazer expressed his opinion to the Swiss board:

Aside from the cracking plant addition, our future principal expansion, I believe, will be in connection with the marketing of our products. For more than a year there has been every indication that we would be forced into a retail marketing of gasoline yet we have been able to postpone such a step without serious detriment to our future. Our present policy is to delay entry into retail marketing as long as we possibly can, at the same time recognizing that such a step eventually is inevitable. An individual case might arise at any time where in order to protect a profitable nearby outlet we might have to make an immediate purchase. Although many of our very good customers have disposed of their businesses to competing refineries, we have

148

during the past year been able to discourage the proposed sale of any of those distributor accounts which we consider absolutely essential to our future distribution. It is my viewpoint that we are primarily in the refining business and that we undertake other lines of endeavor only in so far as it is necessary to protect our refinery investment.

In April, 1930, the purchase of Tri-State Refining Company was an important step in the integration into marketing. Tri-State had been a competitor in the purchase of crude oil from Cumberland Pipe Line Company and had become an important competitor in sales of refined products in the localities in which Ashland marketed gasoline. After the purchase, the annual report for 1930 described the progress in marketing that had been made in the few years of the company's operations:

The gasolines of the two refineries are marketed principally within a radius of 150 miles of the plants in a highly industrialized section of Ohio as well as in West Virginia and Kentucky under the locally well known brand names of Red Pepper Ethyl, Green Pepper Anti-Knock, White Pepper, Tri-State Ethyl, Tri-State Super-Motor, Tri-State Aviation gasoline, etc. A large portion of the Company's refined products is delivered by a fleet of motor trucks operating within a radius of 50 miles. Additional distributing facilities operated by the refining companies and subsidiaries include warehouses with railroad sidings located in ten cities within a radius of 150 miles, oil barges, river-rail terminals and river-pipe line terminals for delivery of gasoline, kerosene and fuel oils. A substantial portion of the gasoline and fuel-oil output of the two refineries is moved in barges to customers located on the Ohio River, thereby effecting a substantial transportation saving.

The refining companies through subsidiary distributing companies own or control more than 50 bulk and service stations. Several hundred independent service stations, not owned nor operated by our refining companies or their subsidiaries are supplied under contract. Several million gallons of refined products are sold annually to major oil companies that have no refineries in this territory to serve economically their retail outlets. . . . In addition to their own compounded motor oils, the refining companies are the exclusive distributors in their marketing territory for both Quaker State and Veedol motor oils.

149

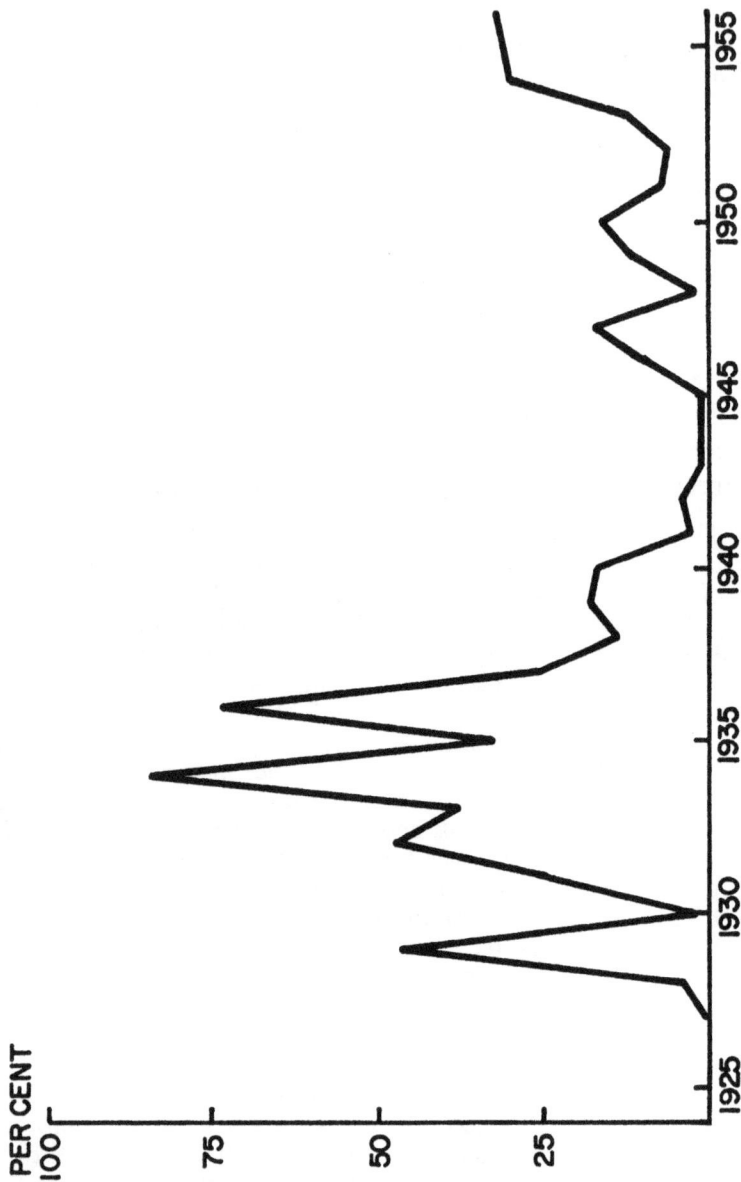

EXHIBIT 17—PERCENTAGE OF TOTAL CAPITAL EXPENDITURES PLACED INTO MARKETING, ASHLAND OIL & REFINING COMPANY, 1924-1956

After the purchases of Tri-State Refining Company and the Cumberland pipeline, Blazer was content to consolidate previous acquisitions and to concentrate on those markets most favorably located for receiving products from Ashland's refineries. The depression caused him to act more cautiously; moreover, it forced greater attention on the marketing branch. During the early 1930's the greatest percentage of capital expenditures was made in marketing (see Exhibit 17). The absolute amount spent in marketing activities also increased.

In 1931 Ashland's primary market had expanded from the immediate vicinity of Ashland, Kentucky, to Columbus, Ohio. By 1936 high rail costs forced the management to retreat from central Ohio. Almost 50 percent of sales volume was being shipped to the Cincinnati area. The shift was determined by the low transportation cost of $.001 per ton-mile which Ashland's new towboat, the *Senator Combs*, made possible.

Exchanges of gasoline with other companies enabled the management to enter additional profitable markets and to maximize the utilization of its water transportation. In the middle of 1933 the company was selling half of its total refinery throughput to other oil companies. Blazer negotiated these sales with personal acquaintances made during his first years in the industry. An illustration of the manner in which these sales and exchanges were interrelated was described to the board in June, 1933, by Blazer:

Since much of the gasoline is being shipped to Cincinnati, I have arranged with Standard Oil of Ohio to give us material from their refinery in the Cincinnati district to fill a portion of our sale to Shell, repaying Standard with gasoline loaded into their barges at Catlettsburg, thereby picking up an additional freight savings of approximately $60 per car. I did not find it easy to make the latter arrangement, since Shell was opposed to taking Standard gasoline and Standard felt the same way about it.

151

Another such sale was to Socony-Vacuum Oil Company through Shell Oil Company facilities from a refinery owned by the Standard Oil Company (Ohio). In such transactions Blazer's ability to act quickly without administrative bottlenecks made it possible for Ashland to gain additional profits during the depression years.

While the company was actually expanding its marketing operations in some areas, it was retrenching in others where competition was intense. Such a fluid marketing program developed into a basic sales policy of the small company. In 1936 Blazer expressed his policy in a quantitative manner to the board of directors:

We hope to give up all gasoline business which is not within ½c per gallon transportation expense from either our plant or the plant of some major oil company with which we can exchange gasoline. . . . We hope that our average will be ¼c per gallon.

The expansion programs of oil companies in the marketing branch caused a serious problem by 1936. Blazer observed in February, 1936, in a speech to the petroleum division of the American Institute of Mining and Metallurgical Engineers:

The largest single item of expense in the oil industry is the cost of marketing gasoline. . . . Even the most casual observer knows that there are far too many service stations. . . . There have been a few half-hearted attempts by members of the industry to justify the present number of gasoline outlets, but the amazingly small average volume of sales per outlet and the high unit cost of sales afford ample evidence to the contrary.

However, a few months later, in August, 1936, Blazer informed the board of directors:

A very substantial part of our expansion this year has been in connection with our distributing facilities. Our new operations are showing up most satisfactorily. Indications are that they will yield a return of between 8 per cent and 10 per cent of our investments, in addition to affording an assured outlet for our refinery. Their value, though, is entirely independent of the refinery and their economic life is not limited by our crude oil supply. Much of the real estate is already worth substantially more than we paid for it.

In spite of his opinion regarding the general problems in the industry, he believed that further attention to marketing was the proper policy for Ashland.

By 1936, pressure was growing in Congress for divorcement of the various phases of the industry. The government was preparing the *Madison* case. In this setting, Blazer testified in 1938 before the Congressional Committee on Petroleum Marketing Divorcement:

The advantages of complete integration of a small oil company, as protection against the alleged threat of monopoly on the part of the larger companies, is especially well illustrated by our own company, since it represents a consolidation of a producing company which, when operating separately, was losing money, an old pipe line which was being operated at a loss, two small refineries which were gradually going broke, and a number of small marketing companies. (The marketing companies, though, were operating profitably.)

Although the company did not expand during 1937 or early 1938, Blazer indicated that he was interested in rounding out its activities. The marketing territory still was located within a radius of 150 miles; it had not changed greatly for a decade. Approximately one-third of the gasoline throughput flowed from company-owned bulk plants. In March, 1938, Blazer expressed his ideas to the board as follows:

Referring further to the expansion we are likely to make, it is not my present thought to engage in any radically different activities, but merely to further round out our marketing operations to include all points to which we have economical transportation costs, to make such additions to our river equipment and river terminals as will reduce transportation costs to present and prospective points of delivery, to modernize our refinery equipment and to purchase such Kentucky producing properties as are offered at a price that should give us a pay-out in not more than five or six years. I believe that these programs can be carried out through the careful use of our depletion and depreciation money. . . . If there should develop an opportunity to merge into our company a relatively small oil company having attractive properties, I would not be averse to the utilization of the 41,010 shares of common stock in our treasury.

A change in refining technology and a change in marketing prospects during the later 1930's caused Blazer some uncertainty as to the desirable amount of integration in marketing. In a letter on April 4, 1940, he said:

There is considerable security afforded for the future by the ownership of assets which have a sound value independent of the fluctuation of the oil industry. I refer to our valuable real estate represented by service stations, etc. . . . Such properties, if judiciously acquired, will yield a fair current return in connection with our business and at the same time insure our having stable assets to help us through any possible period when we might be confronted with losses because of highly competitive conditions within the oil industry.

But ten days later he wrote:

At best there is not much profit to be made in that [marketing] department and I have felt that my time can be spent more profitably in refining operations where much more is at stake. . . . As a whole, the business of our distributing division is profitable, since that department puts out approximately 10,000,000 gallons a year on which the refinery receives an average price of about 1c per gallon more than the average price it received from independent jobbers. The higher price realized from our own distributing division results solely from the fact that our company-owned operations have been so located as to make that business especially profitable because of low transportation costs from our refinery to marketing areas where relatively high price levels prevail.

The latter statement expressed the opinion that was to prevail in the long run. During the 1930's, emphasis had been placed on ownership of bulk plants and on relationships with independent jobbers. Close cooperation with jobbers and deemphasis upon company-owned bulk stations were fundamental marketing policies of the management.

154

Early 1938 sales warranted Blazer's comment to the board: "Never before have we had such a large share of the total gasoline and kerosene business in what we consider our profitable marketing radius." However, by January, 1939, when the gasoline market had weakened, the management gave close attention to the netback of each outlet.

In 1940 Blazer shifted from emphasis upon a small marketing area by opening a new river terminal in the vicinity of Pittsburgh. Shifting of territories and customers continued whenever it became evident that competition was depressing prices in existing areas. Blazer explained: "We are inclined to shift our marketing area farther east, but it will take time to accomplish that result." The latter clause was added in comments to the directors in the late 1930's only after an increase in size of operations. During the early 1930's the smaller size of the company facilitated the use of the flexible policy. There was little evidence in the letters to the directors to indicate that the shifting process previously had taken any appreciable length of time.

The shifting of markets was not the result of a conscious desire to diversify Ashland's marketing operations geographically. In fact, Blazer considered concentration in a marketing area at this time to be an asset. In January, 1942, Standard and Poor's Corporation criticized the localization of Ashland's operations. Blazer reacted to this criticism in a comment to the board:

The fact that our business is localized from a transportation standpoint is its greatest asset. The greatest weaknesses of too many of the large oil companies is that their business is spread out too thinly from a transportation gallonage standpoint.

This localized market was especially valuable during the wartime period. Service to consumers could be offered efficiently by Ashland Oil without aid from the government or from other companies.

155

Rationing of gasoline in the consumer market caused the sale of over half of total throughput to the government by July, 1944. Marketing policies were dictated by government regulation. This made it impossible for the company to continue its flexible marketing techniques. Blazer, spending much time in Washington on official business, found that at the same time he could obtain a better adjustment of his company's marketing program to the governmentally controlled system.

PROMOTION OF AN ADVERTISED BRAND

With the increased refining capacity acquired during the war, Ashland Oil was faced after World War II with a need for accelerating its marketing program. In 1946, sales campaigns emphasized a new brand name, "Ashland"; vigorous sales effort expanded the marketing area of the company; Blazer created a separate marketing division with line authority over all company-owned bulk plants. The annual report for 1946 explained the change required by reconversion:

Reflecting the reversed direction of movement of petroleum products that was required during the war, many of our old customers had been replaced by more distant consumers in the East who could be served only with the aid of wartime governmental transportation allowances or through the exchange of refined products with other oil companies.

With the end of the war and the sudden termination of these arrangements, we soon found it necessary to greatly reduce our refinery thru-put pending a difficult and expensive re-arrangement of our marketing outlets.

During this reconversion period, Blazer developed policies which formed the basis for the expanding marketing program. He reported to the stockholders in 1947:

A large part of the expansion of our sales has resulted from increased volume through our existing outlets. In addition, we have been able

156

to furnish petroleum products to various independent marketers who welcomed the opportunity to become Ashland distributors when their former suppliers, because of shortage of products, encouraged them to seek new sources of supply. Present conditions in the industry are conducive to closer cooperation, on a permanent basis, between independent refiners, such as this Company, and independent distributors who, for many years, have furnished a substantial part of the outlets for nationally advertised brands of petroleum products. Our ability to maintain and increase supplies of refined products for our customers rests, to an important degree upon cooperation between ourselves and the many independent producers of crude oil with whom we are associated. Incidentally, independent oil producers, independent refiners, and independent distributors furnish healthy competition and help to prevent the possibility of control of the oil industry falling into the hands of a few large companies. Products of our Company are sold under the brand-name "Ashland" and also under approximately 150 individual brand-names of independent distributors. Many of the latter use cooperatively the emblem "Independent Quality Service," which identifies them as marketers of products manufactured by our Company.

This emphasis upon the independent marketer was a necessity, since by this time the company marketed less than 10 percent of its output of gasoline through company-operated bulk plants. Ashland Oil owned only ninety-one service stations in 1945, most of which were operated by independent lessees.

Blazer's letters to the board of directors during these years immediately after the war made continual reference to the pressing sales problem:

We believe our markets will not support crude runs much in excess of 15,000 barrels per day which compares with 12,000 barrels per day in 1941. It is not going to be easy to get our overhead down to a proper basis after having built up an organization to do a 30,000 barrel per day business.

· · · · ·

Marketing investments run into substantial sums. I would estimate that we would have to spend at least a million dollars for every additional thousand barrels per day of controlled gasoline distribution. We have an original investment in our marketing facilities of

$1,346,000 (now depreciated to $575,000) through which we distributed an average of 830 barrels of gasoline per day. . . . It is much easier to expand than to contract. We will be able to solve our problems, but the process will continue to be painful.

During this period of reconversion, the company made every effort to acquire additional marketing outlets, resulting in 1947 in the purchase of several large service stations. Blazer advised the board:

Our proposed program for the acquisition of additional marketing outlets is likely to further reduce our earnings, since the cost of each outlet is so high, in most instances, as to commit us to a loss for the first few years while we are depreciating the inflated costs. Nevertheless, marketing facilities are a necessary protection for the operation of the balance of our business.

The increase in refining capacity during World War II formed the foundation upon which Blazer planned to build a more secure operation after the war. The ability to make high-octane gasoline enabled the management to consider an aggressive promotion of the company's own brand name and a spectacular series of mergers.

RAPID EXPANSION OF THE MARKETING BRANCH

By August 9, 1948, Ashland Oil had successfully introduced its new brand, "Ashland," and expanded its marketing outlets. A prospectus summarized the facts:

The major portion of Ashland's output of gasoline, kerosene, and furnace distillates is sold to wholesale distributors who, at their option, market products under their own brand names or the company's established trade name of "Ashland." Somewhat less than 12 per cent of its output of gasoline is distributed through 26 company-operated bulk stations. The company operates tank trucks serving a large number of independent service stations, and 182

158

service stations, owned, leased or otherwise held by the company, but operated for the most part, by independent lessees.

The pressure for markets was always the basic reason from Ashland's viewpoint for encouraging the mergers which took place during 1948-1950. This represented a shift in emphasis on Blazer's part. From 1924 to 1947 he had always showed the greatest interest in refining. After World War II the pressure for markets seems to have caused him to place more and more importance on the marketing phase of operations. At least six major mergers increased the sales potential for the company's products.

In late 1948 Ashland Oil further increased its marketing organization in Kentucky when it purchased marketing properties from Ohio Oil Company. These properties could not be supplied economically from Ohio's refinery at Robinson, Illinois, but they helped Ashland Oil expand its territory in a profitable area.

PROMOTION OF A MULTIBRAND SALES PROGRAM

As a result of the mergers, the marketing phase of the company's operations became more complex. While the policy of selling to those who resold under private brands continued, Ashland Oil now owned at least five brands of gasoline and many brands of lubricating oil. Each of these had become well known in the community in which it had been sold prior to the merger: "Frontier" had first been established as a domestic fuel oil in Buffalo, but later it was accepted as a single-grade brand of gasoline in that area; "Aetna" had developed a strong following in Central, Southern, and Western Kentucky, being sold through company-owned stations; "Freedom" gasoline had built little goodwill, but the "Valvoline" oils had an international reputation; "White Rose" had been a famous name of National Refining Company and was still accepted in a localized market.

After the mergers, Ashland Oil controlled a well-rounded line of products; it had a specialist in heavy residual oils in its Allied organization; it maintained a strong marketing outlet for domestic fuel oil through its Aetna and Frontier business; it marketed several national brands of lubricating oil, including its own brand "Valvoline," and was the distributing agent for "Quaker State" and "Pennzoil"; it kept a reputation for good industrial and railroad lubricating oil through its National Refining organization.

Sales of all products increased during 1950 and 1951 at a rate of approximately 40 percent per annum. Part of this increase resulted from the addition of new facilities which had been acquired through merger. However, in 1951 there were no mergers, and the continued high rate of increase during that year could only have resulted from internal growth of the sales program of the company. To a large extent, the complementary nature of products formerly marketed separately by antecedent companies explains the 1951 increase.

The annual report for 1952 outlined the sales program which then existed:

Although Ashland and its affiliated companies own and serve many hundreds of service stations, the greater part of the Company's output of gasoline and motor oils is distributed through more than six hundred independent jobbers. Ashland endeavors to be the best source of supply for these aggressive, independent businessmen. Frontier, Allied and Aetna divisions deliver heating oils to thousands of homes, principally in Buffalo, Cleveland and Louisville, but, again, independent jobbers distribute most of the company's output of fuel oil for household heating.

Increased sales were closely related to an accelerated program of intensive advertising. Prior to the mergers, Blazer had not felt that the company distributed sufficient gallonage in the area to support an expensive advertising program. Now he commented to the board of directors: "I am impressed with the fact that one of the most obvious disabilities of our group of refining and marketing companies is our failure to develop a cohesive merchandising and advertising program."

In an effort to develop a more unified sales program, the company retained more policy formulation at the home office. It adopted a system of universal credit cards to meet the competition of credit systems of the larger companies. The advertising programs were based primarily on the theme of the company as a local citizen. Suggestions had been received from stockholders that the various brands should be more closely related to the Ashland Oil organization; however, the previous success of the company's local community identification encouraged the management to continue its longstanding policy.

MARKET SHARE AS A MEASURE OF SUCCESS

Not until 1951 did anyone in Ashland's management begin to think in terms of market share. Even at this time the study of market share, from which Exhibits 18 and 19 were extracted, was made by a staff executive who had only recently joined the Ashland Oil organization in one of the mergers. (It is significant that this executive had previously worked for a major oil company.) The charts showed the position of Ashland's sales in the eleven states. Also, it is interesting to note that this study, done after the company had become a major marketer in several of these states, was not made generally available even to the sales executives within the company. In fact, in a number of interviews with top executives in 1954, the knowledge of the market position offered by this study had not been remembered. It was filed away and not used as a basis of marketing decisions during the period after 1951. Figures of the total absolute gallonage and profitability continued to be the measure that Blazer wanted as a basis for his actions.

Even though the study was not used, it does provide a clear picture of the position attained by Ashland Oil in 1951. Exhibit

18 shows the relative importance of the various states in Ashland's markets. Although Kentucky leads Ohio in percentage of total market share by 18.5 percent to 6.8 percent, Ohio leads Kentucky by a wide margin in total gallonage of gasoline sales.

EXHIBIT 18

ESTIMATED PERCENTAGE OF THE GASOLINE MARKET SOLD BY ASHLAND OIL & REFINING COMPANY IN ELEVEN STATES, MAY, 1951

State	Share of Market	Under Company Brands	Under Private Brands	State	Share of Market	Under Company Brands	Under Private Brands
Kentucky	18.5%	15.2%	3.3%	Illinois	0.7%	0.2%	0.5%
West Virginia	7.1	4.8	2.3	New York	0.3	0.3	....
Ohio	6.8	2.8	4.0	Virginia	0.1	0.1	....
Indiana	4.6	1.8	2.8	Michigan	*	....	*
Pennsylvania	2.0	0.7	1.3	Missouri	*	....	*
Tennessee	1.8	0.5	1.3	(*less than 0.1%)			

Source: Independent Market Research Agency's Report to the Company.

EXHIBIT 19

GASOLINE SOLD BY ASHLAND OIL & REFINING COMPANY BY STATES, MAY, 1951

State	Quantity (Million Gallons)	Percentage of Total Gasoline Business	State	Quantity (Million Gallons)	Percentage of Total Gasoline Business
Ohio	15.2	35.7%	Tennessee	1.3	3.1%
Kentucky	10.4	24.5	New York	0.9	2.1
Indiana	6.1	14.4	Michigan	0.1	0.2
Pennsylvania	4.2	9.9	Missouri	0.1	0.2
West Virginia	2.4	5.6	Virginia	0.1	0.1
Illinois	1.7	4.0	Total	42.5*	99.8%

* Does not include sales to other oil companies, sales to government divisions, and consumption by company trucks.

Source: Independent Market Research Agency's Report to the Company.

In Kentucky, Ashland Oil had become the second most important marketer of gasoline behind Standard Oil of Kentucky; in Ohio, the company had become approximately as important as Sun, Pure, Shell, Texaco, Sinclair, and Gulf, the leading marketers in that state behind the Standard Oil Company (Ohio). Other studies do not show clearly the relative importance of Ashland Oil in Ohio, Indiana, and Pennsylvania, since the company sells much of its gasoline under other brands and ships the gasoline from terminals located in other states. In the McLean study, Ashland Oil appears insignificant in Ohio until one notices that between 20 and 25 percent of the total gasoline is classified under "Miscellaneous Marketers."[1] A significant percentage of this "Miscellaneous" and that of several of the smaller listed marketers were actually a part of Ashland's share.

In most of the markets, Ashland Oil obtained a higher percentage in the less concentrated markets than in the urban areas. Sales through company brands were less important than sales through private brands in most states, with only Kentucky as an important exception.

During the period of increase in Ashland's business, the share held by the major companies was declining in four of the important states in which Ashland Oil was operating. Cassady showed that in Kentucky the share held by independents increased from 27.1 percent in 1931 to 33.7 percent in 1949; in Indiana, from 29.5 percent to 38.3 percent; in West Virginia, from 14.7 percent to 16.4 percent; in Illinois, from 27.9 percent to 35.2 percent. In Ohio, the independents' share had actually decreased as a result of aggressive marketing by the Standard Oil Company (Ohio); yet Ashland Oil increased its proportion in this very competitive market to a large percentage of the total business done by independents in the state. In Pennsylvania, Ashland Oil and the other independents were effective only in the section west of the Allegheny mountains; their percentage dropped in the state as a whole from 22.0 in 1931 to 7.5 in 1949.[2]

[1] McLean and Haigh, *The Growth of Integrated Oil Companies*, 106-107. See also E. P. Learned and C. C. Ellsworth, *Gasoline Pricing in Ohio* (Boston, Harvard University Graduate School of Business Administration, 1959).

[2] Ralph Cassady, *Price Making and Price Behavior in the Petroleum Industry* (New Haven, Yale University Press, 1954), 66.

In 1953 the petroleum industry experienced a weak market for its products. Ashland Oil had a large new refinery at Buffalo and faced a decrease in demand with an increase in stocks available for sale. An intense effort was required to increase the number of sales outlets. The statement in the 1953 annual report delineated the company's marketing organization:

> The Company engages in direct marketing operations only when it is unable to obtain adequate distribution through local marketers. Approximately 22 per cent of the Company's gasoline output is distributed through its own local bulk plants, tank trucks, or service stations.

Under depressed conditions, Blazer found it necessary to expand the company-owned marketing operations. The expansion of throughput in the Buffalo area in late 1953 placed unusual pressure on the sales department to market the large amounts of gasoline being produced. The area was one in which independent jobbers were scarce; little could be done, therefore, through established accounts. The time required to expand sales of its own outlets or to establish new outlets was too long to help the short-run problem. By the middle of 1954, gasoline from Buffalo was temporarily flowing back to Cleveland and Detroit. This flow represented a back haul of products produced from crude transported to Buffalo at extra expense.

In August, 1953, just prior to placing the new Buffalo refinery on stream, Blazer predicted the future trend of capital expansion to the board:

> We probably will continue to increase our investment in distributing operations. They will do well if they earn a return of 6 per cent. The most that can be said for such properties is that they are a form of insurance as regards the output of our refineries. Probably we will have more local price wars to contend with; we haven't had much trouble in that respect in the past.

Expansion by merger continued only in the marketing phase. Early in 1954 the company purchased through an exchange of stock Kellogg Petroleum Products, Inc., a large independent distributing company located in Buffalo, to increase company-owned distribution as the market expanded toward the eastern seaboard.

DIVERSIFICATION INTO SPECIALTY PRODUCTS

Emphasis on specialty, high-margin, low-volume products had proved to be a useful marketing policy for the small firm in the 1930's. This policy had been abandoned during World War II because of pressure for volume production, but it was still very attractive to the large firm of the 1950's. Blazer explained in November, 1956:

I am constantly pressing our manufacturing and sales departments to expand our production and sales of industrial and specialty petroleum products where the advantage of public acceptance of trade-marks is unimportant, and where the cost of raw material [crude oil] represents a smaller percentage of the value of the finished product.

By this time the executive attention to these specialty areas and national accounts warranted an additional vice president for marketing. Previous success in asphalt encouraged the management to add asphalt plants at Buffalo and Findlay.

It was not until the middle 1950's that petrochemical production fitted into this attention toward specialties. The Buffalo refinery received the petrochemical manufacturing facilities. In June, 1956, the R. J. Brown Company of St. Louis was merged into Ashland Oil. Wells explained to security analysts its importance and the company's emphasis on specialties:

[It] gave us a very substantial position in the national marketing of naphthas, solvents and chemicals. This company is the largest independent marketer of special petroleum naphthas and solvents. Their marketing area extends from Denver and the Texas-Gulf Coast area on the west to Cleveland, Detroit and Louisville on the east. They purchase the majority of their requirements from many refiners. Their future plans will not disturb their present supply arrangements but will extend their marketing activities eastward and secure additional business in the areas that our refineries can serve most advantageously. R. J. Brown Company will market most of the aromatics and a part of the other products produced in our new plant at Buffalo.

Our Company pioneered in the development of jet fuel and is one of the largest suppliers to the military forces and to engine manufacturers. Also, we were among the first to work with the railroads in producing improved diesel fuel oils and lubricating oils to facilitate their conversion to diesel power. We regularly supply large quantities of diesel fuel oil to more than two dozen railroads.

MULTIPLE MARKETING POLICIES

The policy of selling to jobbers for resale under private brands was retained by Ashland Oil since its very beginning. Since Ashland Oil had no nationally known brand of its own for many years and mergers with other companies provided locally well known brands, Blazer reasoned that it could benefit by keeping the brands with established goodwill. After the merger with Tri-State, the management began to offer more than one company brand. Continued growth through subsequent mergers led to a policy of "multiple marketing."

The term *multiple marketing* refers to emphasis on sales in the same marketing area through more than two channels. As a result of mergers, Ashland Oil found some areas in which sales territories of predecessor companies overlapped. In addition, interest in selling gasoline to owners of private brands increased the variety of outlets. An example will illustrate the unusually large number of channels employed in selling the capacity throughput. In Lexington, Kentucky, the following channels of gasoline distribution were used in 1954. The company owned a bulk plant through which it distributed its "Ashland" brand of gasoline to service stations and consumers. Ash-

166

land Oil was represented also by an independent commission agent who distributed "Aetna" branded gasoline. At least one independent jobber purchased "Ashland" branded gasoline, receiving deliveries by tank car or transport truck direct from an Ashland Oil refinery. In addition, a private brand marketer purchased gasoline from an Ashland Oil refinery and resold the product under his own brand name at a reduced price. Also, the company supplied a number of small independent jobbers who sold gasoline under their own brand names but with an identifying sign, "Independent Quality Service." Blazer had sometimes referred to this multiple marketing policy as "riding all horses in a market."

As the company became larger, conventional marketing techniques were more widely used. Yet, in July, 1955, Blazer explained his reasons for continued attention to multiple marketing:

Offsetting the advertising trends favoring major oil company stations is the fact that such a large number of these stations are being built; each new station dilutes gallonage and adds to the cost of marketing, thereby making them more vulnerable to the competition of the price-sellers which, in many areas, are underselling conventional brands by as much as 5 cents per gallon. . . . In Buffalo, where we have many branded outlets, we are conventional marketers and we have about the same attitude toward the price-marketers as do the major oil companies. In certain other metropolitan areas our policy is just the opposite.

· · · · ·

We have made important purchases from independent jobbers of conventional marketing facilities. In some instances we probably would have been wiser if we had been more active in buying unconventional outlets, some of which have changed hands recently.

GEOGRAPHICAL DIVERSIFICATION OF THE MARKET

In the 1950's Blazer became more concerned with the threat of a low-cost, large-diameter pipeline entering Ashland's marketing territory. Although this concern dated from World War II, he knew that ultimately the company must face the fact that a large line would enter the area. Whereas his policies previously had stressed the economy of a compact marketing area which could be economically served by river transportation, his new policy emphasized geographical diversification. The weakness of concentrating the company's marketing effort in one area became greater.

Early in 1954 Blazer decided to extend the marketing boundaries to any areas in which Ashland Oil could sell products at a profitable netback. This policy of geographical diversification was a second step in the attempt to protect the company from potential threats of new low-cost transportation facilities into its principal territory and to enable it to expand its market during a time in which sales were leveling off. It was reminiscent of the shifting of marketing territories used during the 1930's. This "marketing flexibility" will receive additional attention later as one of the fundamental aspects of Blazer's managerial philosophy.

GREATER SECURITY IN MARKETING OUTLETS

By 1956, insecurity of market outlets again appeared to press Blazer to reappraise the policies which had been so successful for the company when it was small. Ashland Oil had grown to a size which made it necessary to be careful not to do anything which might "rock the boat." The marketing policies that evolved were more conventional. Blazer observed to the board in August, 1956:

Business through our own outlets is more dependable, obviously, than the business of independent jobbers, who may be either raided or bought out by our competitors. I think of our marketing operations much the same as insurance premiums.

The company was neglected consistently by investment analysts because of its lack of company-owned crude oil production and its unconventional marketing policies. In answer to a point made by an analyst that independents sold at lower prices than those of major oil companies, Blazer in July, 1957, explained the status of pricing of the company:

In the case of our Company, our wholesale prices for products compare favorably, I am confident, with those of the major oil companies. About one-third of our gasoline is retailed under our own brand names at conventional prices. A substantial part of the gasoline which we supply to independent distributors for resale under their brand names is sold to the public at reduced prices, but from those independent distributors we receive the same wholesale prices as from distributors who market conventionally at the same prices as the major oil companies and we believe that the average wholesale prices we receive for our products compare favorably with those received by our largest competitors. In fact, it is probable that our company realizes a better return on its investment in refining and marketing facilities than is characteristic of the major oil companies.

By January, 1957, when Blazer retired from the position of chairman of the board of directors, marketing had proved that it was becoming the most strategic branch for executive attention. Ashland Oil still had little crude oil production, but the availability of foreign crude oil relieved worry over shortage of crude in the short run. By 1957 Ashland Oil was accepted as a substantial company in its transportation activities. By 1957 the refining branch had been modernized, and the charge that Ashland Oil owned a number of obsolete refineries was no longer true. It could produce as high an octane gasoline as anyone. The pressure in marketing continued, however, since the company had outgrown the size in which its past marketing policies were most effective. It now was confronted with the simple fact that it needed to standardize its service and to build "prestige" stations which would be comparable with the service stations of the majors.

8

OBJECTIVES OF MANAGEMENT

Concepts of the Role of the Chief Executive

PAUL BLAZER was chief executive of Ashland Oil & Refining Company and its predecessor, Ashland Refining Company, from the incorporation of the latter in 1924 until January, 1957. The refining company was organized only after Blazer had accepted the executive position. The members of the board of directors maintained such confidence in his decisions that he enjoyed complete and relatively unchallenged operating authority throughout the period; consequently, the organizational structure and the managerial processes developed during these years were primarily the result of the opinions and the actions of the chief executive.

This chapter and the next summarize the objectives of the management and introduce various phases of the chief executive's managerial concepts. The tenets of the philosophy are not unusual in the management of a small firm, especially when an executive has grown up with a company and when he has not been exposed to the detailed formal study of management. The doctrine set forth here is somewhat at variance with current thinking of students of management. It is a belief that has developed from observation of the requirements of the situation by a small operator. After the company began to grow rapidly, the chief executive gave increased attention to arguments supporting his views in his letters and oral discussion.

As a competitor, Blazer found that he could operate successfully only if he were able to avoid the disadvantages of the generally approved methods used by the larger companies. Later, this negative approach developed into his "way of doing business." In this period, 1936-1948, intermittent comments in his letters indicated that Blazer gradually was crystallizing in his mind a positive and unified body of managerial concepts. The rapid growth of the company stimulated an expression of his ideas in letters to the board of directors. Opposing views of personnel acquired through mergers had the effect of pointing up disadvantages of his concepts and offered challenges from outside management thought. This caused him to define his ideas more clearly in his efforts to mold all subordinates in his way of thinking.

By the time the firm became relatively large, Blazer was convinced that his concepts had proved successful in the past and would continue to be important in future considerations. He recognized, however, that such principles would be most difficult for another man to follow if suddenly he were to assume the responsibilities of the chief executive. In an interview in 1954, Blazer realized that his ideas would require changes: "The system that I have used will undoubtedly be modified. My successor probably will not occupy the same position as I have. I think that the department heads will instinctively assume more responsibility than they have under me."

Blazer's philosophy never was expressed in any single speech, letter, or article. In fact, certain elements have changed over a period of time and can be understood only through a historical study. In this research a conscious effort has been made to refrain from reading reasons into certain statements or actions. No attempt has been made to change the concepts so that they might appear more logical. If the managerial process is partly nonlogical, the objective analyst should not attempt to establish a framework by the help of hindsight. In a dynamic situation, it is quite possible that consistency through time is not a virtue. As conditions change, the philosophy may be supplemented in order to fit the newer problems.

To organize these elements into a unified group of concepts, an attempt was made to establish certain useful generalizations which were supported by past actions. Upon review of these statements, Blazer qualified them and indicated that he preferred to avoid generalizations. His attitude was summed up in two sentences: "I have not been as interested in the rules as I have been in the exceptions to the rules," and "I have great abhorrence of fixed patterns." He expressed a similar idea in his letters to the board when he stated that if he had not acted unconventionally in the past, the operations probably would not have resulted in so great a success.

McLean and Haigh have pointed out that the small refiner has developed unique areas of operations and products in order to fill the "interstices" of the industry. They explain the success of the smaller companies in niches which have been overlooked by the large companies.[1] The management of Ashland Oil has emphasized this idea of "interstices" not only in problems of an operational nature, discussed in previous chapters, but also in the problems of managerial techniques.

Discussion of Blazer's philosophy will relate to his relationships with the stockholders and the board of directors, and with the executive personnel immediately subordinate to him. The former encompass the general economic, social, and political philosophy fundamental to the orientation of the top management and its place in the larger structure of the economy. The latter which include the managerial concepts fundamental to the manner in which the operational policies have been executed, are explained in Chapter 9.

[1] McLean and Haigh, *The Growth of Integrated Oil Companies*, 632.

Any study of top management should treat the fundamental ideals and goals upon which the operational policies are based. In this sphere the scientific search for facts must give way to the ethical search for values. Barnard's scheme of the moral element directs one's attention to the long-run endurance of the organization—important for outlining the frames of reference in which the short-run, expedient, and opportunistic approach has predominated.[2]

A competitor of Ashland Oil expressed this ethical aspect in different words: "The success of Ashland has been the result more of the vision of the chief executive than from any other factor." The term "vision" in this statement embodied an idea of that ability of a top executive to think in terms of the long run while directing action in the present.

In Ashland Oil & Refining Company there is no question that Blazer established the objectives of the firm. The legal theory that the board of directors determines the fundamental objectives is only partially satisfied. Ultimate approval was a function of the board, but the initiation of ideas was a function of the chief executive; consistently the board supported his recommendations.

Several factors enabled Blazer to maintain this position. In the early period he alone of the top executives was familiar with the operating aspects of the industry, but he disseminated the most important facts to individual members of the board. Moreover, he repeatedly contacted individual members by letter and telephone to get their opinions on specific matters. While a single member of the board had no legal authority except as a part of the group, his opinions, determined prior to the meetings, became of fundamental importance in the decision-making process. Although Blazer's prestige in the organization made it unnecessary for him to seek information and opinions so fre-

[2] Chester I. Barnard, *The Functions of the Executive* (Cambridge, Harvard University Press, 1938), 282.

quently, this technique enabled him to advance to the board as a group recommendations which he knew would be approved because of his previous contacts with individual members. Each member felt that he had already expressed himself and therefore was inclined to support the proposals of the chief executive. Although the casual observer usually considered Blazer to be a one-man top management, the indirect impact of the members of the board on the formulation of company objectives should not be underestimated.

Throughout the history of the company the majority of the board of directors were "outside members." During the early period of Swiss Oil Corporation, the president, general manager, and general counsel were operating executives on a board composed of eleven members. The most important group on the board during the first twenty years were the Chicago financial interests who had become stockholders in 1919. A second important group represented the investing public of Kentucky. As a result of mergers during the later years, former coowners of antecedent companies constituted a third important part of a board that ultimately was increased in size to seventeen members (and then reduced to fifteen).

Blazer's monthly letters performed the function of keeping these different groups of the board informed. The board of directors was a professional group with ultimate control dispersed to several balancing interests. The success of the company prevented any of the minority groups from being able to exert control, and so the authority remained in the chief executive in spite of his lack of ownership of a large amount of common stock. In all cases he reported unfavorable factors in great detail. Thus, he usually achieved results more favorable than his forecasts had indicated, thereby minimizing the pressure of unfavorable developments. Accounting records of the company usually reflected this conservative outlook.

A chief executive of a corporation usually has some difficulty in obtaining the maximum help from his board of directors.

174

Much has been written presenting the arguments for an "inside" board and for an "outside" board. Gordon and others have made provocative remarks about the entire institution of boards as the top level of management.[3] In practice, the board often receives orders from the chief executive rather than, according to the basic theory of private property and the corporation, delegating authority and responsibility as representatives of the stockholders.

Blazer, as chief executive, succeeded in keeping close touch with his board, the majority of whom were "outside" members, and he recognized the directors to be the source of new ideas and action. Even though he was not a majority stockholder, he did not run into any potential group who would even think about a revolt in management. The key to this achievement was the constant communication with each member of the board. The telephone was used very intensively. The basic facts and thoughts, however, were transmitted by means of monthly letters, quotations from which have been used throughout this book in support of facts concerning the decision making during the 33-year period. Blazer's first letters were two or three pages in length, but in the 1950's they became longer and often included communications from other executives and reproductions of press releases and other material, so that frequently they ran to forty pages.

The following letter is provided as a sample of Blazer's letter-writing technique. It was selected for several reasons: it was of typical length; it was written in December, 1956, one month prior to the change in chairmen of the board; it was written at a time in which the critical Suez oil situation provided facts which the reader can easily remember; it summarizes several of the ideas mentioned in other sections of this book; and it provides all of the important text of one letter rather than excerpts from several reports, such as have been used in other parts of this book.

[3] Robert A. Gordon, *Business Leadership in the Large Corporation* (Washington, Brookings Institution, 1945), 343-51.

Our annual report is now on the presses and should be ready for mailing by the 20th of the month. I was late in completing my work in connection with the report, and in order to meet our schedule I found it necessary to accompany Arthur Points [controller] and Orin Atkins [executive assistant] to Chicago the early part of last week to read proof and make such changes in copy as always are necessary at the last minute to balance out pages and pictures to obtain satisfactory appearance. From Chicago, Orin and I went direct to New York where we had considerable business to transact—also, a dinner of the Asphalt Institute which was attended by Hugh Jenks [asphalt division manager] and Mr. and Mrs. George Beddoe [asphalt engineer] of our organization.

I have been extremely busy working with Everett [Wells, executive vice president] and others in our organization to adjust our operations to take full advantage of changed conditions resulting from disturbances in the Middle East. Temporarily we are refining at Catlettsburg almost 60 per cent more crude oil than before we put into operation the new equipment only a few weeks ago. Friday and Saturday we averaged more than 70,000 barrels of crude oil in the No. 2 plant where we are endeavoring to test out maximum capacity. It is my recollection that formerly we considered that plant as having a capacity of 35,000 barrels per day with an additional 20,000 barrels in the No. 1 plant, giving an over-all theoretical capacity of 55,000 barrels per day. I believe we never succeeded reaching an average of 53,000 barrels per day for any month except last January when we refined almost 56,000 barrels per day.

In view of the figures mentioned, I am sure that each of our directors will appreciate the significance of refining 70,000 barrels in a single day in the No. 2 plant along with a potential of at least 15,000 barrels per day in the No. 1 plant. The latter plant has actually processed well over 20,000 barrels per day for a considerable period of time, but I presume its capacity would have to be reduced some when running so much in the No. 2 plant since some of the facilities of the two plants are joint—such as gasoline treating capacity, shipping facilities, etc. With such a big step-up in refinery throughput in such a short period, there are many associated problems such as getting that much crude into the refinery and moving out so much products. We have succeeded in increasing, to an estimated 70,000 barrels per day, the amount of crude oil that we can get to the lower Ohio River from the oil fields of Illinois, Indiana, and Western Kentucky. From that point to the refinery

176

much additional barge capacity has been required. Similarly, extra barges are necessary to move the refined products, since almost all of the additional products have to move by water.

We don't have a market for the amount of refined products we are now producing, but we have considerable empty storage capacity. We have had a burdensome inventory of crude oil, so we have been glad to convert some of it into refined products. Besides, we recognized that there was a possibility that, in order to supply additional crude oil for Europe, the Federal Government might seize excess stocks of crude oil (such as we have had) and make them available to major oil companies who are short of crude oil for their affiliated European refineries.

We have considerable flexibility in our transportation system, so we have been able to move the extra oil into our refinery and are doing equally well in getting refined products out to terminals. Also, we have been able to take on much additional burning oil business, since there is a shortage of such oils on the Gulf Coast which, ordinarily, supplies a part of the requirements of our marketing area.

Our increased production helps Allied Oil Company which appears to be able to use more residual fuel oil than it can obtain from the refineries which historically have supplied a substantial part of its requirements. However, as indicated earlier in this letter, our principal purpose for pressing for all of this extra production at this time has been to ascertain our maximum capacity and locate the "bottlenecks." In some places we find we need larger pumps, which usually can be obtained by transferring from some other service where there is extra capacity. Also, we always have on hand spare pumps. We are not making substantial expenditures at this time in order to remove bottlenecks which are showing up at these high levels of throughput, but we are obtaining valuable data as to what we would have to do in order to maintain this high throughput regularly, or to increase it to a predetermined level such as might be desirable if our country should be involved in serious hostilities, or if, for some other reason, our volume of sales should increase unexpectedly. Besides, there is great value in having some extra capacity which relieves some of the pressure of having to engage in overtime maintenance work whenever something breaks down. For years we have had so much business in relation to our refining capacity that we couldn't afford to lose a single day's production unnecessarily.

I suspect that before we complete these tests we will have refined in a single day in our various refineries more than 160,000 barrels of oil, not including Niles which could be brought into operation on short notice.

We averaged last year less than 125,000 barrels of crude oil daily. I hope we will have sufficient market to support 140,000 barrels per day for the current fiscal year. We don't have that much business in sight at this time. In fact, we did not draw on inventories as much as I had anticipated in October when we ran only 115,129 barrels per day because Canton was down for inspection and the new equipment at Catlettsburg was not in full production.

The incremental cost of refining additional barrels of crude is very low. Thus, any increase in refinery throughput is extremely profitable, unless in order to sell the extra production we have to absorb excessive transportation charges to move our products into a wider radius. Of course, we have an expanding demand in our normal marketing area, but there is a limit as to how big a percentage we can take without competitive problems which result in price disturbances. Probably one of the most important aspects of our business is the ability to properly appraise such situations.

The time to expand our business for burning oils is like the present when there is a shortage. When our competitors are short of supplies they offer less resistance to giving up a few of their customers. We are selling considerable material to major oil companies. We can hardly expect to retain that business under normal conditions, although the changed pattern of supply might enable us to transfer some of our current sales to exchange arrangements, which would enable us, economically, to get into new, more distant markets.

Since nearly all refiners now have an excess of gasoline, we have not put as much pressure recently on obtaining new gasoline customers as in expanding our burning oil sales. We are meeting that situation, primarily, by reducing our yield of gasoline obtained from a barrel of crude oil and increasing our yield of burning oils. In that manner we have been able to step up our crude oil runs substantially without making much more gasoline, but under that program we not only have to sell the burning oils from a larger number of barrels of crude oil but also a larger percentage yield of burning oils.

Generally speaking, a smaller percentage yield of gasoline is likely to mean less realization from each barrel of oil refined and, thus, a smaller profit. However, under present conditions the spread between burning oil prices and gasoline prices is not great and to a large degree the difference is offset by the lower manufacturing cost which has to be charged against the incremental barrels of crude oil refined.

The program we are pursuing currently obviously puts a lot of pressure on our sales department and, also, to almost as great a degree on our various transportation facilities and on our crude oil supply department. Even within the last week we have committed ourselves to purchase additional crude oil over periods up to a year. We probably will continue to have more crude oil than we require, but in view of the situation in the Middle East we are assuming that if we find we have too much crude oil we should be able to sell our excess.

Our directors should be very appreciative and proud of the way our department heads and their assistants respond to the extra demands associated with the program I have outlined; it has some of the characteristics of a "fire drill," except it requires considerable time—probably a few weeks.

Walker Marx [Canton refinery manager] has been able to further step up the throughput of our Canton refinery and has a throughput slightly in excess of 30,000 barrels per day. Buffalo could run almost as much as Canton except that the Buckeye Pipe Line is able to deliver only slightly over 26,000 barrels per day. The Buckeye people indicated to me in New York last week that it should not be many months until they can give us approximately 29,000 barrels per day. We had been expecting only 28,000, which is on the basis of our receiving half of their anticipated 4,000 barrel per day increase, but Buckeye says that the Socony Mobil plant at Buffalo is not showing much interest in taking more than 1,000 barrels per day of the increase. I am unable to prophesy as to whether Buckeye would give us the extra thousand barrels if Socony should change its mind. Heretofore we have been on a 50-50 basis.

Enclosed is our latest forecast of working capital and capital expenditures. We endeavor to be conservative in our forecast of earnings; I imagine we have considerably underestimated our earnings for the fall and winter quarters. It will be noted that our October earnings were at the rate of more than $4,000,000 per quarter. Of course, there are a great many current uncertainties, but we probably won't fall far short of a profit of $8,000,000 for the first half of our fiscal year. Much depends on the weather which will greatly influence our sales of burning oils and, thus, determine the level of our refinery throughput. We are now estimating our capital expenditures at approximately $100,000 more than is indicated by the forecast.

An increase in the price of crude oil is generally expected. I presume it is being held back by current oversupplies of gasoline and the weakness of that market. A shortage of crude oil—especially in the hands of independent refiners—should strengthen the gasoline market.

From a long-range standpoint, a crude oil advance probably would be unfavorable to us. Temporarily it probably would cause us to show increased earnings, since presumably prices for refined products would advance sufficiently to absorb the crude oil advance and we would pick up some profits on our inventories. At the time of the last advance, refined products prices gradually eased off after the initial advance and at the end of a year were about as low as previous to the advance and in some instances lower.

Enclosed is copy of a letter sent last week with employee-dividend checks. Attached to it was a copy of my letter to stockholders reporting earnings for the fiscal year and advising of recommended changes in the management of the Company. Originally we sent the employee-dividend checks quarterly, but after their size declined so much (one quarter the payments had been equal to 11.53% of wages) we have been sending single checks in December for the fiscal year ended September 30. I believe that everyone connected with the management of our Company is of the opinion that this profit-sharing plan is justified from the standpoint of the Company. In that connection, I am attaching to the letter which was sent to employees some figures which were compiled for me by Jim Barker [chief accountant] showing the payments by years since the plan was organized nine years ago.

Our annual meeting falls on January 21. We will call a directors meeting for Saturday, the 19th, and I think it would be desirable for as many directors as can do so conveniently to arrive in Ashland on the 18th. Weather permitting, one of our planes probably will cover the Cleveland-Pittsburgh-Buffalo-Canton area. Another plane should cover Louisville and Lexington. Eastern Airlines has good service from Chicago into Ashland via Louisville, leaving Chicago at 3:05 P. M. (C.S.T.) and reaching the Ashland-Huntington Airport at 6:55 P. M. (E.S.T.). I understand that Earl Weaver [executive assistant] and Miss Riddle [a secretary] will be handling transportation arrangements, checking anticipated arrival times, local reservations, etc.

There appears to have been an entirely satisfactory reaction to the proposed management changes as was suggested in my recent letter to stockholders. As far as I can tell, the proposed changes met with general approval within our organization. In fact, I think no one was surprised. Everyone has the utmost confidence in Rex [Blazer, new chairman of the board] and Everett [Wells, new president].

Survival was the primary objective of the management in the early period. Although large profits were made in the first few years, Blazer refrained from expanding at too great a rate. The conservative credit policy, the retention of earnings, and the continued use of relatively old equipment were his reactions to the earlier policy of Swiss Oil Corporation. Probably Blazer reasoned that first he must build security within the structure of Ashland Refining Company. For example, in his report to the Swiss board in 1925, he stated:

In spite of the fact that these improvements would pay for themselves in a few months, it seemed in view of small profits not advisable to make the expenditures in question. . . . If present profits continue as we believe they will, we shall not forget how quickly refining conditions can change, and with that in view we shall continue to keep down our expenditures and endeavor to increase our cash reserves.

His early success in the development of a good credit rating for Ashland, so that the subsidiary could obtain funds at times when Swiss could not, illustrates the result of this drive for survival first.

After survival seemed assured, the primary objective changed to gaining greater security of the company's competitive position. In this connection, growth became important. While growth in itself never appeared to be an objective, numerous comments pertained to expansion as a means of "insurance." Integration into phases other than refining was looked upon as a means of protecting refining operations. Preceding chapters have summarized the successive steps in this integration process. In the early period Blazer appeared to be concerned over the problems of becoming large enough to compete efficiently; in the later years he seemed to fear that the company might become too large.

Beginning in 1936, Blazer made it clear that growth in itself was a poor objective. His speech in 1936 to the American Institute of Mining and Metallurgical Engineers in New York City expressed this view:

In recent years almost every oil company in the country seems to have been willing to sacrifice transportation economies and marketing efficiency for the mere sake of growing larger and covering more territory. . . . Unfortunately, the sales departments usually have been "gallonage minded" which is the antithesis of "profit-minded."

In 1939, speaking before the National Petroleum Association, he stated that the smaller firms should look to their advantages and avoid the problems of large size:

Let me again state that, in my opinion, small enterprises in the oil industry have few problems that arise inherently from their own size, but that their greatest problems, as well as those of the large companies, result directly from the fact that many of their big competitors which are already too large from the standpoint of public policy, and also too large to function efficiently, are striving to become still larger. . . . Actually the earnings report of those companies would look a lot better at the end of the year, if the heads of those companies would tell their respective sales departments that their output for the coming year would be reduced and that it would be necessary to get rid of their least profitable 10 per cent of business.

Later, during the period in which Ashland Oil was growing rapidly, Blazer continually repeated such statements as: "I consider, however, that there is no virtue in size as such" (1950) and "these large expenditures are not made merely for the purpose of having a larger company" (1951). The progressive steps in rapid growth were taken in spite of the fact that they made a larger company. The management felt that the indivisibilities of technological equipment, the need for numerous specialists in the accounting, legal, technical, and other areas, and the instability of the refining sector of the industry forced greater size even though largeness had important disadvantages.

Another objective of the management was the classical maximization of profits for its owners. Although a major proportion of the owners were not operators of the company, they remained so closely related to affairs of the company that the management necessarily looked to net income as a most important objective. For Blazer this objective was based upon pride in the efficient management which profits would indicate, rather than upon the mere acquisition of greater income in itself. His incentive was primarily one of "doing a good job" for men whom he knew personally and who were depending upon his executive actions. The result was that attention to net profits was probably as great as if the company had been owner-operated.

A third objective of the management related to responsibility to employees and to the public. The development of the "Ashland family" illustrates a means by which the management sought to gain its goal. Location of the firm in a small community caused the management to identify its goals with those of the locality. This objective, which involved treating each employee on a personal basis, had definite effects upon techniques of management.

The family goal of the management established an atmosphere that permitted the development of strong informal organizations. While these spontaneous groups often were evident only upon close study of the operations, they formed a most important aspect of the total organizational picture.

This objective extended to all those with whom contacts were made—the community, business suppliers, customers, and so on. Blazer believed that good public relations could best be achieved by every operating employee without formalization of functions. As a result, no executive held the title of public relations specialist; rather an atmosphere of informal action was promoted with the conviction that more sincere, and therefore more effective, public relations would result.

Varied methods have been used in striving toward this family objective. Large Christmas parties have been held each year since 1930; in recent years 12,000 people were entertained. Informal actions by individual executives have met many emergency needs of employees. For example, upon the death of the father of a junior executive, a company car was delivered personally by the vice president to facilitate a necessary 150-mile trip. On another occasion Blazer provided his personal oxygen equipment to help an employee's wife who had a serious heart condition.

As Ashland Oil expanded through the merger route, it became necessary to decentralize the personnel function and to stress employees' identification with their former companies. Certain past owners were retained in operations and the previous names of the merged companies were kept, even though legally they became operating divisions of Ashland Oil.

Many factors, however, have caused difficulty in the maintenance of this family concept. Unionization in 1933 laid the foundation for a division of loyalties of employees; government regulations during the 1930's and the World War II period necessitated formal policies rather than reliance on personal discretion. The mere increase in the number of personal relationships prevented executives from maintaining contact with each person and his individual problems.

As the company became larger, the management initiated several actions in an effort to maintain high morale by means of the family concept. In line with the theory held by Blazer that it would be advantageous for the employees to own the stock of their company, the company offered several stock purchase plans to employees, beginning in 1938. A type of profit-sharing plan was initiated in 1947 in which "employee dividends" were paid quarterly to employees and declared by the board of directors in the same manner that the board declared dividends on stock.

A fourth objective of the management was to promote competition in the industry through positive actions and to rely upon restrictive legislation such as antitrust laws only in exceptional cases. Although Blazer managed a small company which was the type that the government sought to help, he consistently fought efforts to freeze a formal pattern by regulation of the industry. This attitude first became clear during the days of the NRA codes in the 1930's, when he was an important industry leader on government committees. This experience with government administration had a lasting effect upon his ideas of management. Many of the formal techniques employed in government did not fit administrative experiences in the small company. In December, 1933, he first indicated his thinking to the board of directors: "Our company . . . has less to gain and more to fear than others from this regimentation of the industry." It was his belief that Ashland's advantages were based upon a unique group of factors which would be neutralized in any governmental effort to "stabilize" or "regulate" the industry. A small growing company should not seek to freeze a pattern in which it has a small share. This attitude persisted after Ashland Oil grew to large size, as indicated by Blazer's speech to the Ohio Petroleum Marketers Association, Cincinnati, September 20, 1950: "Federal control and regulation may be forced upon us because of our own lack of leadership, but I can't understand why any individual or group should voluntarily ask for such disaster." It is interesting to note that Ashland's return on investment was lowest in 1934 and 1942-1945, both periods of greatest governmental controls (see Exhibit 4).

Throughout the 33-year period, Blazer took an aggressive part in the fight of the "independents" against the "majors"; yet he preferred to emphasize direct competitive means rather than appeals to Congress or to the courts. In 1938 Blazer wrote to a fellow independent,

I want you to know that we consider your problems a matter of concern to independent companies in this territory, since in your

185

desperate efforts to save your investments you may rock the boat and we will all sink. . . . And when the independents pass out of the picture we are certain to have either government control or ownership of the oil industry as prophesied a few days ago by Assistant Attorney General Thurman Arnold. I believe that instead of attacking the present advantages of the major oil companies, all independents should try to obtain competing facilities. The problem can be solved only by sound economics and not through politics.

In fact, Blazer gave testimony before Congressional committees which had the effect of supporting the interests of the major companies. Example of such testimony is that offered to the House Interstate and Foreign Commerce Committee, July 14, 1953, in which Blazer explained the factors involved in price increases of crude oil in support of the position of the integrated producers, even though the increase proved actually detrimental to the interests of Ashland Oil.

Several times this action was taken to show the majors the importance of the support of independents—a part of the larger picture of "bargaining power" with the larger companies. Such moral suasion to get an "even break" from the larger companies —through admittance into pipelines, through exchanges of gasoline, and so forth—were in Blazer's opinion much more effective than appeals to legislative or judicial power. It was evident throughout the letters and interviews supporting this study that a most important place in the decision-making process was given by the chief executive to what David G. Moore called "external economic strategies."[4] Moore's term, *strategy*, is especially descriptive of this important characteristic of Blazer's philosophy.

The mere fact that Ashland Oil has continued to be classified as an independent has had an effect upon the aggressiveness of its relations with the government. It has continued to seek favorable action in ways in which other companies of approximately the same size feel they cannot. Major companies in the past

---

[4] David G. Moore, "Managerial Strategies and Organization Dynamics in Sears Retailing" (dissertation, University of Chicago, 1954).

have been confronted with unfavorable public opinion of their action; consequently, they remain passive on issues at times when Ashland Oil aggressively leads.

The managerial ideas and policies employed by Blazer are often criticized on the grounds that they do not tend to result in the development of new executives to take over the responsibilities upon retirement of the original organizer. As events turned out, Blazer did have the opportunity to develop new blood. One of the chief concerns of the financial experts in the early 1950's was what would happen to Ashland Oil if Paul Blazer had to drop out? Persistent rumors, even as late as 1955, were that some major company would buy out Ashland Oil. It still was unbelievable that the company could continue to prosper without the help from one of the major companies. The thought by many in the investment market that Ashland Oil could not exist after the retirement of Paul Blazer caused much speculation.

Actually, Blazer had always given great attention to training of subordinates on how to make a critical analysis for a good decision. It is true that the final decision was usually made by Blazer, but the type of executive which developed under the Blazer approach can only be judged by the results after retirement. Blazer's letter to a member of the board of directors in July, 1957, five months after a new set of officers had been in operation, gives a clue to the state of management which existed after the "strong" man retired:

I am glad that you were favorably impressed with the way Rex [Blazer, chairman of the board] and Everett [Wells, president] handled the recent meeting. I am just as impressed with the way in which they are handling their other responsibilities. I consider that we made the proper decision in the manner in which we divided the management of the Company between them. Crude oil production,

187

crude oil supply, pipelines, river transportation, refining, wholesale marketing, accounting and part of the activities of the Personnel Department are under Everett. Allied Oil, Cleveland and Lakeland Tankers, our bulk plants and service stations, sales of Valvoline and our other branded products, advertising, most of our public relations and much of our personnel relations (other than labor negotiations) come under Rex. They both appear to be happy and I am confident that neither of them will disappoint us.

I am not conscious of having had any difficulty in meshing my activities into theirs. I consider that I have very few direct responsibilities, except in connection with the budget and financial matters. I have more time to read and plan. Having no administrative duties, I spend more time with Everett and Rex than formerly, although only in an advisory capacity. I continue to do considerable work for Everett as regards efficiency of refinery operations, coordination of crude oil supply, transportation, refinery production and sales. I have more time than he to dig up information, but I am careful to let him make the final decisions. Both Everett and Rex recognize that they have full responsibility for all decisions, that they are not obliged to take my advice on anything, and that if any mistakes are made they must assume the responsibility.

In my opinion, the new responsibilities of Rex and Everett are no more significant than the greater reliance of all of us on our department heads. Although some of them are quite young, they have been developing rapidly and are capable of carrying much greater responsibilities than a few years ago.

I am especially glad that you are favorably impressed with the present situation, since it is highly important to the morale of Everett and Rex and their assistants that our directors show their confidence in them. I am not aware of any significant mistakes, but if some are made it is important that everyone recognize that new management is entitled to make a few.

We have an excellent team with many years ahead of them before normal dates of retirement. Everett is 52 and Rex will soon be 50. Palmer [Talbutt, vice president, sales] is 58, and his very capable office assistant Joe Davis is 41. John Fred Williams [vice president, industrial relations] is 52, Chub Moffitt [vice president, national sales accounts] is 42, and Bob Yancey [vice president, refineries] is 36. Erskine Owens [pipelines and crude oil supply manager] is 46. He has an excellent assistant in Jack McClure who is only 31. Orin Atkins, who is capable of being part of the top management, is 33.

Ned Seaton [treasurer] is within three years of retirement and considerable thought is being given to the selection of his successor. Alex Chamberlain [vice president] who serves as a trouble shooter and usually has time for special assignments, is 57. Upon the death of Mike Dupree [transportation manager] his responsibilities were split up between John Fox, Bob Gray, and Dave Beldon all of whom are 39, and Bill Meachem who is 40. Ward Disbrow who heads supply and distribution is 40, and his excellent assistant Paul Kinnaird is only 34.

   . . . . .

I doubt if there are many companies of our size in which the chief responsibilities rest on men of so young an age.

9

FLEXIBILITY AND INFORMALITY

Blazer's Concepts of Business Management

UNDERLYING THE concepts and practices of management which evolved in Blazer's mind, the idea of flexibility has been pervasive and fundamental. Flexibility in this context refers to the quality of adaptability to external changes, susceptibility to modification of actions, resiliency of policies, and responsiveness of the entire organization to meet new problems; it is the antonym of rigidity, unchangeability, inflexibility, and inelasticity.

The idea has been that a small growing company in Ashland's position in an expanding industry requires adaptability in place of the advantages of the technical concept of bureaucracy important to the large firm. Dimock defines bureaucracy in its rigorous sense as having three requisites: subdivision of jurisdiction, hierarchy, and professionalism of personnel.[1] Max Weber and other sociologists have expanded the study of this subject as interest in the organization of large-scale governmental and business units has increased. Robert K. Merton has stated the advantages of its consideration in the management of a large

[1] Marshall Dimock and H. K. Hyde, *Bureaucracy and Trusteeship in Large Corporations* (Temporary National Economic Committee *Monograph No. 11,* Washington, 1940).

unit: "The chief merit of bureaucracy is its technical efficiency, with a premium placed on precision, speed, expert control, continuity, discretion, and optimal returns on input. The structure is one which approaches the complete elimination of personalized relationships and non-rational considerations (hostility, anxiety, affectual involvements)."[2] Bureaucracy, therefore, is antithetical to the doctrine under study.

Only in the latter part of the 33-year period has Ashland Oil been a large firm in any sense of the word. Its chief executive has believed that concepts of bureaucracy were best applied in the largest firms. He would agree with Robinson, who points out: "Where the important decisions are infrequent, and the necessity for a quick decision is less urgent, the large firm can play its part more efficiently."[3] Blazer has felt that in the refining branch of the petroleum industry quick decisions on important matters occur frequently.

Dimock has identified the nature of the large firm and has focused attention on the contrast between problems of the small and large organizations:

Large corporations, like other large human enterprises, are bureaucratic. . . . Organization grows in importance as size increases and trusteeship gains ascendency. And like other large organisms, the larger the modern corporation becomes, the more it tends to move slowly, adapt itself with increasing difficulty, be increasingly concerned with its inner rules and procedures. Hence, it stands in danger of losing that flexibility of price adjustment and resiliency of managerial outlook which is the most valuable social asset of free competition.

. . . . .

The extensive use of more or less rigid and precise rules and working procedures is well-nigh universal among the giant corporations. . . . Within a corporation, furthermore, subordinate officials often complain that they are not permitted sufficient discretion, that trans-

2 Robert K. Merton, "Bureaucratic Structure and Personality," *Reader in Bureaucracy* (Glencoe, Illinois, Free Press, 1952), 363.
3 E. A. G. Robinson, *The Structure of Competitive Industry* (New York, Harcourt, Brace and Company, 1932), 50.

actions of any size must be referred through too elaborate channels. When questioned on this, however, the superior officials are quick to point out that there are aspects of such transactions which subordinates cannot understand. For consistency and safety, therefore, the top executives claim that limitations are necessary. Indeed, a whole philosophy of management—that of scientific management—is largely predicated on the idea of elaborate rules established for the most simple and routine actions. An unwieldy body of regulations is thus inevitably built up, and it becomes easy to forget the objectives of the system in a blind obedience to rules.[4]

These quotations have referred to the term *bureaucracy* in a technical and rigorous manner with no idea of an opprobrious meaning. The term also has been used popularly in a derogatory sense. Von Mises believes that the characteristics of a bureaucratic organization will not appear in a private business operated for profit. His use of the term is more limited than Weber's and Merton's, but he makes several statements which are consistent with Blazer's understanding of the shortcomings of bureaucracy: "Bureaucrats . . . are no longer eager to deal with each case to the best of their abilities. . . . Their main concern is to comply with the rules and regulations, no matter whether they are reasonable or contrary to what was intended."[5]

These concepts of bureaucracy relate to management of large firms; they are the antithesis of flexibility as it applies to the smaller firm. The recognition of this concept of flexibility is by no means new; many writers have referred to certain aspects of the idea. None, however, appear to have placed as great an emphasis upon its application to a growing firm as did the chief executive of the Ashland Oil & Refining Company.

A major part of the concepts of management that proved useful to this management can be classified under five distinct types of flexibility: technological, marketing, financial, personnel, and organizational.

[4] Dimock and Hyde, 3-4, 33.
[5] Ludwig von Mises, *Bureaucracy* (New Haven, Yale University Press, 1944), 41.

The term *technological flexibility* pertains to Blazer's concept of maintaining the mechanical ability to change one's equipment to produce those goods desired by the consumer when and where he wants them. In refining, it involves the maintenance of specialized equipment in standby condition ready to make a product in limited amounts. In transportation, it involves the physical ability to change sources of raw materials where they can be most economically obtained and to shift refined products to those locations where competitors are placing less pressure on prices. The idea of technological flexibility is the element of the management's concepts best defined in Blazer's letters since the very beginning of operations in 1924. His report to the Swiss board of directors in 1928 indicated the concept in refining: "Such an arrangement [pipe stills in place of shell stills] frankly could be operated much more economically than our present plant, but it would not give us the flexibility that we now have, nor the special products which add so much to our profits." Larger companies were installing pipe stills; shell stills had become obsolete. In spite of this trend in technological improvements for volume production, Blazer preferred to keep shell stills (in addition to newly constructed pipe stills) in operation until 1938 because of their adaptability to production of high-margin, small-volume business. Restricted financial resources necessitated that he maintain profits without installing expensive equipment.

Blazer gave an important place to flexibility of refining equipment throughout the critical depression years, as has been shown previously. By 1945 mechanical refining flexibility proved valuable in another way. For years the company had depended primarily upon eastern Kentucky crude. The necessity to adapt refining operations to other crudes was explained in a company publication:

The equipment of this refinery has been designed to operate on a flexible basis. It is capable of refining crude oils of widely varying

193

characteristics ranging from the low-gravity, high sulphur oils from certain fields in Texas, Mississippi, and Western Kentucky, through medium-gravity, low-sulphur content crudes from Eastern Kentucky, Illinois, and other mid-continent fields, to high gravity, distillate oils from producing fields adjacent to the lower Mississippi River.

Refining flexibility involved both ability to use various types of crudes and ability to turn out various types of refined products.

One of the best examples of this flexibility appeared in 1949, when the price of residual fuel oil decreased by 50 percent. The annual report of 1949 stated:

There was no compensating decline in the cost of crude oil. Fortunately, the flexibility of Ashland's principal refinery at Catlettsburg was again demonstrated when the yield of fuel oil from a barrel of crude oil was reduced to approximately 2% during the period of greatest oversupply. As a result Ashland's earnings from its refining operations suffered much less than those of the oil industry in general.

Previous discussion of the company's transportation policies has made it clear that barges permitted greater physical flexibility; boats could be shifted within limits of navigable rivers, whereas pipelines remained fixed from one source to one destination. Probably Ashland's reason for using barges was its limited financial capacity; however, continual change in routes of towboats is evidence that this phase of transportation flexibility was no small factor in the operations of the company. The annual report for 1948 pointed out this idea:

For a company which specializes in refining, the constant shifting of the economics of the oil industry necessitates transportation flexibility in relation to sources of crude oil supply and markets for refined products, as well as the ability to vary yields of the different products from a barrel of crude oil. Our principal refinery, located near Ashland, Kentucky, enjoys the flexibility of cheap transportation for crude oil and products via the Ohio River which is intersected by pipe lines from both the Eastern Seaboard and the Southwest. Likewise, this refinery has exceptional flexibility in its ability to maximize or minimize the yield of any of the more important refined petroleum products.

A second element of the concept of flexibility has been predominant in the marketing of refined products. The term *marketing flexibility* applies to the concept evident in the techniques of this chief executive that distribution of refined products should take whatever channels, pricing, and methods that conditions on the spot seem to dictate. Examples of shifting geographical marketing areas have been evident in previous discussions. Generally the move was eastward to the "economic ridge." Early sales in Columbus, Ohio, decreased and sales in Cincinnati increased as the economics of transportation caused changes in netbacks. In 1940 the shift to Pittsburgh resulted from intense competition in other areas and favorable costs along the Ohio River. In 1954 the policy of shifting marketing areas to obtain geographical diversification was considered by Blazer to be important to the ability of the company to make profits.

The policy of primary dependence upon the independent jobber was an integral part of marketing flexibility. Smaller amounts of company funds were needed for marketing properties; the local community served as the primary factor of sales promotion in place of national advertising; the company was not chained to any market as it might have been if bulk plants had been owned by the company. The idea of the efficiency of the independent oil marketer remained a cornerstone of operating policies. Blazer reasoned in a speech in May, 1950:

Many independents have made small fortunes in the oil industry. I believe there are just as many opportunities for the independents in the future as at any time in the past. In fact, some of the trends in business today are working to the advantage of individual enterprise as contrasted with large scale operations. Tax laws which

195

already penalize larger companies will be even more favorable for small business in the future. Labor relations are likely to be better in smaller operations. . . . It is becoming increasingly difficult for large corporations to retain the friendly attitude of their employees in the face of propaganda directed at large employers and big business in general. . . . All that the independents need is an "even break."

Independence of price-making action was illustrated by a member of the board of directors of one of the "major" companies in whose "territory" Ashland Oil operated. A contract for a large volume of business went to Ashland Oil in a city where the major company had an efficient refinery. Upon discussion of this matter by the board of the major, it was recognized that the reason that Ashland Oil could enter into the "front yard of the major's territory," even though it had no refinery in that city, was that Ashland Oil had little other such business in the community and could base its bid solely on the price that would yield it sufficient netback to its closest refinery. The major was unable to meet this price without destroying its price pattern and stability in its primary marketing area.

Ashland Oil was able to use its refining flexibility to a greater advantage than could larger companies because its established commitments of any given type of refined product were not as definitely fixed. Ralph Cassady commented on this idea, "It takes very little, if any, increase in price to bring forth additional quantities of product. . . . Usually all that is needed, particularly for major companies, is orders for goods from customers. It should be pointed out, however, that quantities of one type of product (such as gasoline) may be increased at the expense of some other type (heating oil, say) by independent refiners when the relationship between the two favors such a move. This type of flexibility may not be enjoyed as much by major companies as by minor companies because commitments to customers rather than price advantage have a more important bearing on what is to be produced."[6]

[6] Cassady, *Price Making and Price Behavior in the Petroleum Industry,* 18-19.

Growth had its effect in those areas in which acceptance for the company's branded products had been obtained; there were signs that the sales policy gradually was adjusted to encourage stability in those territories in which Ashland Oil distributed a larger proportion of gasoline. These areas, Louisville and Buffalo, for example, became increasingly important in the total sales volume. In 1954 Blazer expressed a feeling that probably he had encouraged the sales of private brands to too great an extent in some major areas. Several "price marketers" placed pressure on the price structure of areas highly important to Ashland's own branded sales.

The policy of selling to all types of outlets continued after the increase in scale of operations in spite of the fact that the company acquired several important brands and became more intensely integrated in marketing. The primary reason was that refining capacity had increased at a greater rate than sales outlets and the company found it desirable to utilize the "multiple marketing" policy to move refined products. In May, 1950, Blazer restated his previous ideas relating to price marketers in a speech:

I think the oil industry makes a mistake in thinking disparagingly of so-called price-cutters who try to undersell the market. Within reasonable limits, they are good for the industry. We shouldn't "look down our noses" at them. I suspect they are our best protection against governmental investigation. They tend to keep our prices down and force us to be more efficient. Besides, the most ardent price-cutter today may be a most respected member of the industry tomorrow.

The policy of selling to distributors for rebranding has required marketing flexibility. Adaptability to several reference marketers encouraged further emphasis on flexibility. Cassady makes the following observations:

The principal characteristic of price making in the minor firm is that decisions can be made without reference to committees or other executives. This flexibility results, in part at least, from the fact that this type of concern is definitely the follower type and also that the organizational structures of such concerns are very simple and authority and responsibility are concentrated in a few key individuals. Very often in such organizations the vice-president in charge of sales makes most of the pricing decisions, although at times the president wishes to be informed on such important matters as any serious soft market conditions in which the company is involved or any unusual types of deals which the firm has taken on and especially those which involve financing or financial obligations.

. . . . .

This flexibility is one of the most valuable assets of this type of organization.[7]

When Ashland was small, flexibility was mandatory for survival in the particular niche in which the company operated. Flexibility made possible a dynamic outlook that encouraged rapid growth. After achieving large size, the emphasis on branded products to some extent opposed the flexibility demanded previously.

FINANCIAL FLEXIBILITY

The term *financial flexibility* refers to the ability of a firm to change its investments from one type of activity to another and to secure additional funds quickly on good terms when an opportunity appears for profitable investment.

As a result of the stringent times experienced by Swiss Oil Corporation, the management of Ashland Refining Company watched its financial condition carefully throughout its existence. A policy of relating capital expenditures to the amount of allowance for depreciation was used as a rule of thumb to avoid financial difficulty. By 1928 (when an 8-percent preferred stock was issued) the subsidiary was better able to obtain new funds than its parent. Listing on the New York Curb Exchange of

[7] Cassady, 104.

the securities of Ashland Oil & Refining Company in 1936 was a further step in gaining increased flexibility of financial actions.

In 1945 the company became known in the national securities markets upon sale of 40,000 shares of 4¼-percent cumulative convertible preferred stock in April and the refunding of an insurance term loan by issuance of $5,000,000 of twenty-year 3-percent sinking fund debenture bonds in August. The working arrangements with an underwriting firm, developed in these two issues, laid the foundation for numerous entrances into the securities market in the next ten years. Ashland's good financial name cannot be overestimated in its effect on the growth of the company. One observer commented that he felt that Blazer had not considered rapid growth until the success of his entry into the national investments market in 1945.

Each time the management sought new funds after 1945, the amount depended upon an estimation of the maximum which the company could obtain on good terms. Funds "for working capital" were obtained when such funds were available in the market rather than when they were needed. It was assumed that the funds could be used profitably at some time in the future, since the industry was an expanding one. Evidence of this action does not appear in consecutive balance sheets because, by the time the money was actually obtained, the management usually was able to put most of it to work before the date of the next annual report. Comparison of current ratios of Ashland Oil with those of other companies shows no significant difference, partly because the management did not give attention to the current ratio but to net working capital. Blazer reasoned that the use of net working capital for general management planning was preferable; unlike the current ratio, it was not susceptible to being improved merely by more rapid payment of the company's current obligations.

References to this policy of securing funds whenever possible appeared from time to time in Blazer's monthly letters. An example of one such statement occurred in October, 1939, during negotiations for a loan from an insurance company: "We do not need the money at this time. . . . Possibly we will not make the anticipated large capital expenditures before two years, but in the meantime we can use much of the proceeds."

A capital budget was not formally prepared until 1950, and then only as a flexible guide for department heads, when a serious illness incapacitated Blazer. However, even then the budget was more an estimate of source and application of funds than it was a capital plan. Blazer late in 1951 stated that "in an opportunist business like the oil business, budget estimates do not mean much." In spite of this statement, a budget committee was appointed and depended upon for its estimates.

Until the late 1930's the method of deciding upon capital expenditures used postponability as a criterion.[8] This method kept the management cognizant of the obsolescence factor so important in the refining branch. As long as the company could refrain from committing itself on new capital expenditures, it retained the ability to shift its future operations to meet new technological advances. Committing only a minimum of its funds to fixed assets at any one time gave the management that much more reserve power to meet new situations. The small refining firm has such limited financial resources relative to the size of a capital commitment, that it must continually maintain its ability to shift to possible new courses of action. The introduction of the Dubbs process in the 1920's illustrates the adaptability in refining; the discovery of new fields in Illinois in the late 1930's was an example in production.

PERSONNEL FLEXIBILITY

The term *personnel flexibility* refers to the ability of the executives and employees of the company to perform in various posi-

---

[8] For further discussion on this method, see Joel Dean, *Managerial Economics* (New York, Prentice-Hall, Inc., 1951), 567.

tions within the organization as they may be needed. This fourth aspect of flexibility relates to dependence upon executives who are "generalists" rather than specialists. Blazer felt that this phase of flexibility resulted in high morale and good cooperation and was, therefore, essential to other aspects of his philosophy. He outlined this idea in a commencement speech in 1950:

The demand for trained men to fill the best jobs in large companies exceeds the supply. There are more vacancies near the top than down the line. I don't know why that should be, unless it is because large business is inclined to make specialists out of its key people, thereby precluding breadth of experience. Possibly a contributing factor is the necessity for the various levels of authority in big business to make decisions and assume responsibility for those under them. Or it may be because so much emphasis has been placed on technical training and not enough on those qualities which make for leadership. In that connection, we must never forget that essentially we shall always live in a world of people—rather than in a world of things. Others can do for us more than we can do for ourselves.

He considered a training program tailored to the needs of a particular person as part of the answer to the problem. His executive development procedure was handled on an informal basis to accomplish this. He explained that in many cases large companies have had to secure their top executives from small companies because the large companies had developed only specialists. A company in the oil industry is especially troubled with this problem, since its four branches are very different from one another and require men of different skills and backgrounds. The small oil company must encourage executives to look at integrated groups of problems; it cannot afford to maintain specialists who do not see the problems of other departments.

The small company must usually develop its own top executives. In only a few instances has Ashland's management gone outside for experienced executive personnel. The company started with young college-trained men and gradually developed a team of executives who became accustomed to working together. After its rapid growth through merger, it relied upon the executive talent of the newly acquired companies to supplement its supply of experienced executives. It maintained a low overhead with little "fat" in order to keep costs low. Moreover, it increased the breadth of ability of those executives it did have.

A contrasting statement was made by the officer of a large company: "If you want to develop management, you need to have some 'inventory' to work with, some people to rove around, some men 'in transit', some extra jobs where men are in training for greater responsibilities. You must have spares."[9]

A second aspect of personnel flexibility involves implementation of personnel policies. Even in 1954, when the company was relatively large, there was no "employees' handbook" nor any precisely stated group of personnel policies (except those which were required in government regulations and union contracts). The emphasis continued to be upon personal and individual relationships between management and employees even though the company was one of the first to recognize the Oil Workers International (CIO). The contest for loyalty of the workers between the management and the union has been strenuous. According to a union leader, often after union officers had attempted to achieve unity of effort on the part of employees, Blazer through personal contact would convince the individual workers that their best interest called for their support of management.

ORGANIZATIONAL FLEXIBILITY

The foundation upon which all other aspects of flexibility have been built will be referred to as *organizational flexibility*. The concept refers in part to that "quality in an organization that

---

[9] Frank W. Pierce, *Executive Talent: Its Importance and Development* (California Institute of Technology *Industrial Relations Bulletin 19*, Pasadena, March, 1951).

enables it to adjust itself to temporary changes in business conditions without serious losses of economy or effectiveness."[10] The concept also refers to dependence on "informal organizations" and a minimum formal structure—evolution in place of planned engineering. It differs from the concept of personnel flexibility in that the latter pertains to the human element, whereas the former refers to structure and processes.

Although it would be possible to obtain degrees of flexibility in other aspects without special emphasis on this element, Blazer's idea was to avoid a static pattern of structure for fear that it would defeat his efforts to maintain the adaptability of his total managerial effort. Since little thought was given to organization, the emphasis on organizational flexibility was mostly unconscious for a number of years.

In spite of his attempt to keep a clear and simple line structure, by the late 1930's Blazer was forced by the increased size of operations to depend on functionalization (a personnel department was organized in 1938). Later, after the rapid growth of the late 1940's, he began to decentralize some authority in sales divisions to retain organizational flexibility. Throughout, he increased his own span of control to an extreme degree.

In the early period there was little need for an organization chart. When the company became larger, Blazer remained adamant in his opposition to the use of charts. Reference to this idea did not appear in his letters to the board until the firm became large (about 1948). It was at this point that he rationalized from his past experience that conventional organization charts create too much rigidity to fit his other ideas of flexibility. Thus, not until the growing company demanded more formal structure did Blazer begin to organize his arguments against it.

Continued emphasis on these five types of flexibility promoted efforts to achieve the company's objectives. Blazer believed that

[10] Ralph C. Davis, *The Fundamentals of Top Management* (New York, Harper & Brothers, 1951), 508.

his doctrine uniquely fitted the needs throughout most of the company's history. Environmental and historical factors previously discussed provide a basis for understanding this opinion. In later years the skeleton of these managerial concepts appeared in writing; however, for this study it was necessary to supplement the written statements with extended interviews. Tape recordings and on-the-spot observations form the basis for the following discussion.

The elements of Blazer's managerial concepts can be classified into concepts of organizational structure, concepts of organizational process and internal communications, and concepts relating to the evolution of responsibilities.

CONCEPTS OF ORGANIZATIONAL STRUCTURE

During the early years Blazer dealt with his operating problems directly through personal leadership and few subordinates. Previous discussion of refining growth has shown that new functions were assumed by assistants. Although there was some functionalization, Blazer preferred the "staff assistant" to the "staff specialist."[11] Most of the young and inexperienced junior executives required training, and delegation of authority necessarily was minimized. The company was of such a size that one bad mistake by an inexperienced executive could wreck it; Blazer felt that he just could not afford to allow new officers to "learn by their mistakes." Any academic criticism of this breach of currently accepted principles of management should consider the environment in which this technique was used.

As the company grew and required more managerial personnel,

---

[11] Ernest Dale, *Planning and Developing the Company Organization Structure* (American Management Association *Research Report No. 20*, New York, 1952), 61. "The 'assistant to' has no power to act on his own. Instead he furnishes his chief with information and recommendations which the latter is free to use as he pleases. . . . The position of 'assistant to' may have the advantage of introducing a new function 'under the wing' of the boss. Thus the incumbent of the new function is given a chance to show what he can do."

Blazer tailored his organization to the personal characteristics of his subordinates and his needs for assistance. There was no attempt to establish "positions" with definite titles or clearly defined duties. In the early 1930's organization planning first took on importance. Observation of organization disadvantage experienced by other companies caused Blazer to take a definite stand against the use of organization charts in his company. In spite of advice from new executives, external pressures to establish a clear-cut structure, and additional pressures resulting from increased size, the chief executive continued to oppose formal organization planning.

A series of charts could have been used to explain the organization as it appeared at a series of points in time. In this case, however, it would be misleading to use this conventional tool which shows the flow of authority in the usual lines, because it would mean the addition of a synthetic tool that was not actually used in practice. Moreover, this study focuses its attention only on the relationships of officers who normally had direct access to the chief executive. For these reasons, the chart in Exhibit 20 has been developed, principally from interviews, to summarize the actual evolution of the organizational structure of major functions directly under the chief executive.

Exhibit 20 shows a number of characteristics of the top management organization and the personnel that filled the positions over the 33-year period. There were relatively few positions and executives in the first four as compared with the last four years; between 1924 and 1928 only one line and three staff executives answered to Blazer; in 1952-1956, there were at least a dozen line and a dozen staff and functional executives who were normally in direct contact with the chief executive. Many additional executives on the lower levels might have received direct telephone calls or visits from the chief requesting information or giving orders.

The exhibit also shows the gradual development of organization from a single executive performing line functions, to the development of departmentation in the early 1930's, to the beginning of functionalization in the late thirties, and finally to the development of operating divisions after 1948. Moreover, it pictures the development of specialization of functions.

As the company increased in scale of operations and number of personnel, the emphasis upon a loose structure at the top contrasted with the more clearly defined organization of lower echelons. This difference became even more marked after the mergers of the 1948-1950 period. In spite of the known objections of Blazer to charts, many lower executives planned their departments more or less privately with the use of charts and directories. This difference between the levels with respect to formality will further explain Ashland's organization. First, the reasoning behind Blazer's philosophy only partially permeated to the lower echelons. Second, the lower levels were staffed with younger personnel, schooled in the advantages of charting and formality. Their techniques were the product of institutional training in business administration instead of the gradual development of a *modus operandi* acquired from experience. Third, the more routine and standardized problems of the lower echelons lent themselves more easily to formality. Major executives had worked together over a long period of time and knew the management's "way of doing business"; this made the loose structure of the upper levels possible. The low turnover of executive personnel over the entire history of the company prevented any major crisis in the top working group; on the other hand, new personnel were continually added to the lower levels.

By the time the company had reached larger size, Blazer had formulated his reasons supporting his ideas on organization. He recognized that there was greater difficulty in the application of the concepts in the larger organization and that some other person in his position would not be able to carry out such a degree of informality. In an interview he observed:

My form of organization was better adapted to a company somewhat smaller than we are today and in an industry less complex. This type is only practical where the management has grown up with the company. I don't think anyone could come in cold.

The emphasis has remained on company rather than functional departmental loyalty. A central idea in Blazer's concepts was expressed in an interview:

If you had really sharp departmental lines, you would find people telling others that "it is none of your business" and "leave that to me." . . . Nobody is going to take that attitude if they don't know themselves the limits of their responsibilities. I have felt that you get more cooperation from people if your organization is so set up where they have to cooperate to get along.

Immediately after the mergers, special effort was made to integrate the new employees into the "Ashland family" while at the same time making use of their loyalty to predecessor companies. The emphasis was on identification with each of the former company groups but a continual guard was kept against functional segmentation of loyalties. Good communications were maintained between these divisions and the home office. In spite of this effort to build an unsegmented identification with the company, the increased size, geographical decentralization, and different historical backgrounds of antecedent companies made this technique increasingly difficult. The result was the use of the decentralized division for the organization of each merged company.

The closest approach to a visual pattern of the organization was first compiled on December 1, 1951, in the form of a directory, which was distributed to major executives only. Although it included the names of major and middle executives and their position titles, often the individual himself did not know the exact title of his position as stated in the directory. No titles of positions nor names of departments appeared on the doors of the home office. When questioned on the possible significance of this omission, Blazer indicated that he had not consciously planned such a practice. Such incidental evidence, however, is

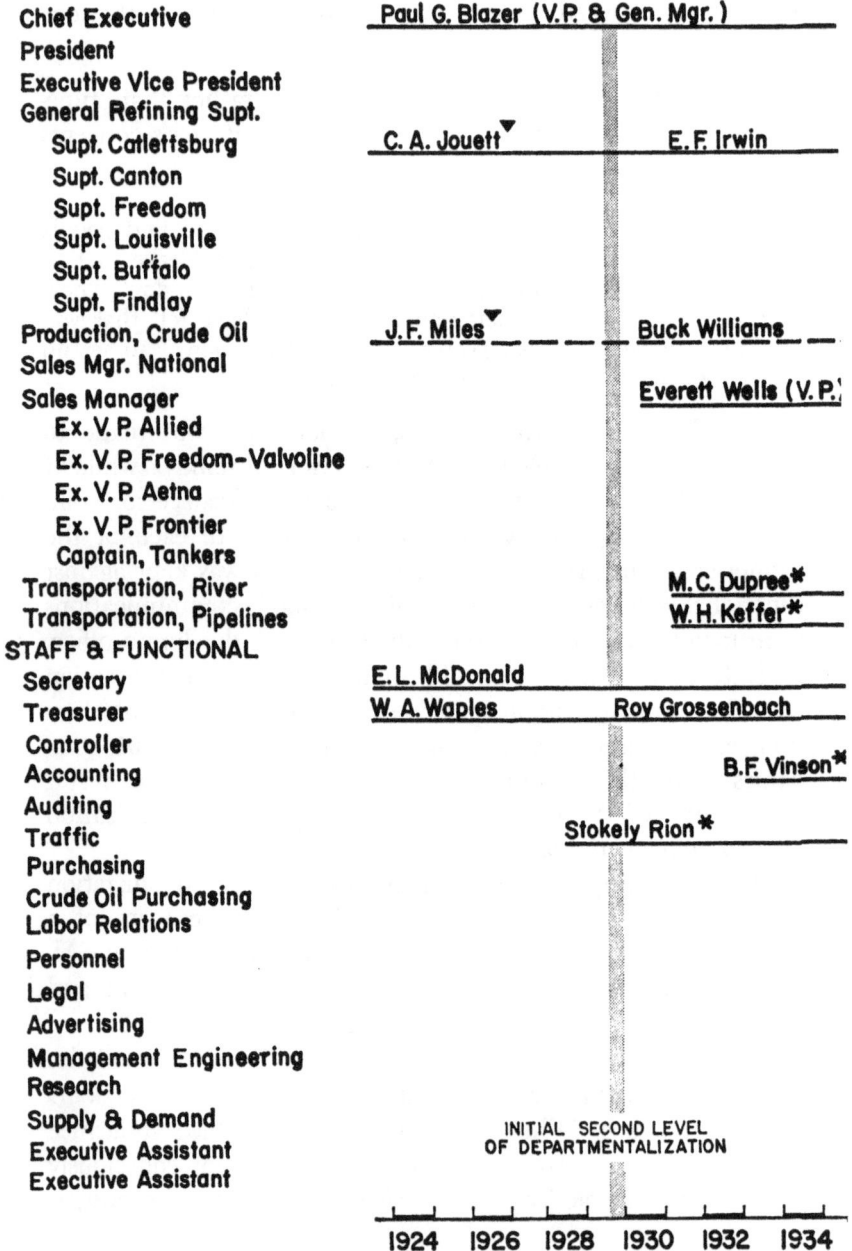

Exhibit 20—Evolution of Organization,

LINE

Chief Executive	Paul G. Blazer (V.P. & Gen. Mgr.)		
President			
Executive Vice President			
General Refining Supt.			
Supt. Catlettsburg	C. A. Jouett▼		E. F. Irwin
Supt. Canton			
Supt. Freedom			
Supt. Louisville			
Supt. Buffalo			
Supt. Findlay			
Production, Crude Oil	J.F. Miles▼		Buck Williams
Sales Mgr. National			
Sales Manager			Everett Wells (V. P.)
Ex. V. P. Allied			
Ex. V. P. Freedom-Valvoline			
Ex. V. P. Aetna			
Ex. V. P. Frontier			
Captain, Tankers			
Transportation, River			M. C. Dupree*
Transportation, Pipelines			W. H. Keffer*

STAFF & FUNCTIONAL

Secretary	E. L. McDonald		
Treasurer	W. A. Waples	Roy Grossenbach	
Controller			
Accounting			B. F. Vinson*
Auditing			
Traffic	Stokely Rion*		
Purchasing			
Crude Oil Purchasing			
Labor Relations			
Personnel			
Legal			
Advertising			
Management Engineering			
Research			
Supply & Demand			
Executive Assistant	INITIAL SECOND LEVEL OF DEPARTMENTALIZATION		
Executive Assistant			

1924	1926	1928	1930	1932	1934

Paul G. Blazer (President)   Paul G. Blazer (Ch. of Board)

J. Howard Marshall▼   Rexford S. Blazer

Everett Wells

E. A. Brown▼   R. A. Whealy   R. Yancey (VP)

B. Morgan   L. Alexander   W. Marx   G. Myers

W. B. Borst   W. Marx

B. L. Heath   J. Blount

E. E. Van Buren

M. C. Sumpter

Cecil West

W. H. Keffer✳ (V.P.)   H. C. Arnold C. Marshall▼ R. Ralston▼ E. Emrick

F. M. Moffitt (V. P.)

P. Talbutt (V.P.)

F. C. Colegrove

H. I. Johnston

C. M. Alexander⏺   F. Ware

W. D. Baskett

L. M. Jonassen

R. Gray

Ray Laughner

E. W. Seaton

A. J. Points

E. Emrick   A. J. Points   J. H. Barker

S. E. Hippensteele

P. H. Banks⏺   W. Meachem   C. Loving

Edward Emrick   J. K. Barron

W. M. Luthy▼   E. Owens

John Fred Williams (V. P.)

Alex Chamberlain   Lewis Ware

S. M. Burnam

Willis Munro✳   G. Sisler

Alex Chamberlain (V. P.)

H. C. Houlton

W. Disbrow P. Kinnaird

INITIAL   INITIAL   W. Hull   O. Atkins
FUNCTIONALIZATION   DIVISIONALIZATION

E. Weaver

| 56 | 1938 | 1940 | 1942 | 1944 | 1946 | 1948 | 1950 | 1952 | 1954 | 1956 |

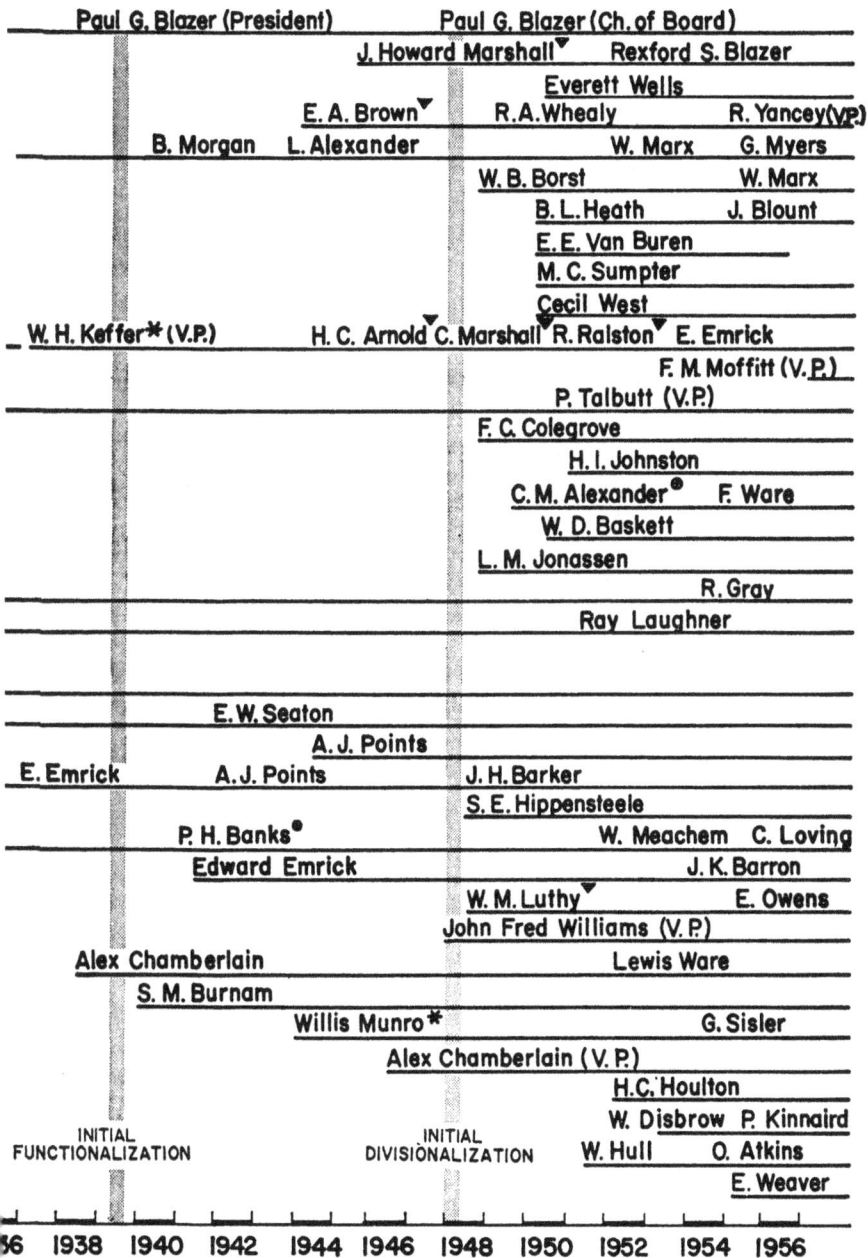

▼Left Company   ✳Died   ⏺Retired

significant in showing the pattern of his entire managerial emphasis.

The layout and furnishing of the offices did not indicate any effort to show the rank of executives through the use of prestige symbols. Both junior and senior executives of the company (except when addressing Blazer) called each other by first names.

Little use was made of a formal status system. The previous discussion of the dynamics of the industry and particularly of the operations of Ashland Oil within the industry has pointed out the changing conditions that confronted the management. Barnard stresses stability as important in the development of a status system:

> The overvaluation of the apparatus of communication and administration is opposed to leadership and the development of leaders. It opposes leadership whose function is to promote appropriate adjustment of ends and means to new environmental conditions, because it opposes change either of status in general or of established procedures and habitual routine.
>
> . . . . .
>
> From what has been presented, it is perhaps evident that the sum effect of the status system, though essential to coherence, coordination, and *esprit de corps*, is to reduce flexibility and adaptability. When the external conditions to which an organization must be adapted are stable, the importance of flexibility and adaptability is much less than under rapidly changing conditions.[12]

Blazer looked upon any definite position title as potentially a straitjacket. For example, if he allowed an executive to become identified with some segment of the organization through clear delegation, he felt that he would be outlining a bailiwick in which that executive would feel a vested interest and would tend to oppose a change which would destroy his delegated departmental "empire."

Throughout the first thirty years of the company the number

---

[12] Chester I. Barnard, "Functions and Pathology of Status Systems in Formal Organization," *Organization and Management* (Cambridge, Harvard University Press, 1948), 240, 242.

of vice presidents was kept to a minimum. The title of vice president did not identify a chief functional executive as such; it referred to the recognition of general ability. No predetermined group of duties carried the prestige of the title of vice president, because different phases of operations alternated in receiving major emphasis. Rank depended upon the importance of the job in current operations, not upon a title structure fixed in some past period. In 1954 there were four vice presidents: only two held line authority—the executive vice president and the sales manager; the other two held staff positions— the director of personnel and an industrial engineer. As the company became consolidated as a large organization, additional vice presidents were appointed in the last two years of Blazer's administration.

Flexibility of titles also permitted Blazer to make the maximum use of the individual support from the former owners of predecessor companies. After the Freedom-Valvoline merger, one owner was placed on the board of directors of the parent company; the other served as chairman of the board of Freedom-Valvoline. In the case of the Allied owners, both were placed on the board of directors. In the case of Aetna, the surviving owner remained in the position of president of the subsidiary. One owner of Frontier retired, and the other became a member of the board of directors.

Precise job descriptions of executive positions were deliberately unstated. Blazer felt very strongly that the degree of cooperation that had been attained in the Ashland Oil organization had resulted from this reluctance to compartmentalize any part of the operations. He felt that numerous disadvantages of such compartmentalization greatly outweighed its advantages. For example: if there were a clear-cut, continually emphasized set of duties for each person, there could be a tendency for the executive to avoid any other responsibilities that might arise in a dynamic situation; emphasis on rigid channels and departments would confuse organization means with organization goals; a spirit of teamwork in the company as a whole could not develop as easily when specialized departments and ranks were emphasized.

Blazer always believed that a "flat" organization was preferable to an organization with a number of levels in the hierarchy. He placed emphasis upon short lines of authority. He felt that it was more economical to be "underorganized" than "overorganized." Administrative overhead remained low as a result of the following reasoning, quoted from an interview:

Our organization started out being very simple. Probably we have been underorganized rather than overorganized. I always said that one man may have a job, running it successfully, and not be overworked. He may be running the whole department by himself, or he may be running two departments—and happy in his work. You give him an assistant. The job may not have changed a bit; at the end of a year, he couldn't possibly do the job himself; he becomes dependent on the assistant. If you gave the two a third man, they would become dependent upon him. That is true when you start organizing and delegating responsibilities. It is better to let one man, in my opinion, cover a little too much territory than it is to have two men. Two things happen. One is that they begin to get jealous of each other—each starts to feel that the other is getting out of his bailiwick. On the other hand, if one man is doing the job, he is carrying it pretty well in his own head; he doesn't have to spend a good share of his time conferring with someone else and coordinating what each is doing. You put two men on the job; you can't expect to get twice as much work. They have to confer a lot to keep from getting their wires crossed; they do get them crossed to a certain extent and spend time getting them straightened out again. Now the weakness of my philosophy is that there is a tendency to be short-handed organizationwise.

In an analysis of the past titles of the major executives, two terms appear to have special prominence and significance. As the firm grew rapidly after World War II, Blazer appointed several "executive assistants" (or assistants to) and "coordinators." These vague titles appeared not from a lack of a clear understanding of the duties of the position, as is often the case, but from Blazer's basic concept of organizational flexibility and general executive development as opposed to specialists' development. Later, in the exposition of the evolution of responsibilities, this idea is shown to be a positive action by the chief rather than mere irrational procedure.

CONCEPTS OF INTERNAL COMMUNICATIONS

The organizational structure evolved at Ashland Oil depended upon communications that would help Blazer to respond quickly to new conditions. Two classifications are important in considering the flow of internal communications. Vertical communications relate to lines of authority from top to bottom of the hierarchy. Horizontal communications relate to means of informal cooperation among executives on the same level of the structure. Henri Fayol characterized these two classifications as ladder and gangplank concepts.[13]

Since Ashland's organizational structure was often not well defined, the lines of authority were not clear. In the early years, authority descended in a simple line from the chief executive. As the firm increased in size, additional young executives were appointed as semiline officers. Often they functioned as staff executives, but as they became more experienced, they began to operate more with true line authority.

Direct vertical communications were encouraged. Blazer felt that he operated not only as a decision maker but as a "reporter" among the various departments.[14] He encouraged the collection of complete information and its dissemination to all subordinate levels which needed the facts. The greater the number of levels through which this information passed, the greater was the opportunity that the information would become inaccurate

13 Henri Fayol, *General and Industrial Management* (New York, Pitman Publishing Corporation, 1949), 34-35.
14 Blazer's experience as a reporter on his father's newspaper during his adolescence and on his college paper possibly affected his attitude toward this function of a manager.

because of the human tendency to add impressions to original facts. As a result, Blazer contacted the person closest to the information, although that person might not be his immediate subordinate.

While all quantitative reports were distributed in writing, instructions for future action usually were oral. The need of the smaller company to change its plans quickly was thought to be met better by a minimum of memoranda. The very fact that a policy was in writing would cause the subordinate executives to rely on the letter of the policy in place of understanding its intent. Blazer felt that once a policy was placed in writing, there was a tendency for it to remain in the files long past the time of its applicability.

Gardner and Moore express Blazer's feeling nicely:

There is a strong tendency for many organizations to develop procedures for handling practically every problem situation—to define everything to the point of eliminating all individual decision and initiative. In considering many modern industrial organizations, it would seem almost as though they were being slowly strangled by an over-elaboration of procedures and policies. In some organizations, no one dares to breathe unless he first finds out what the procedure is. Everything has been defined. Nothing has been left to the imagination or individual decision. Rigid, unadaptive bureaucracy reigns supreme. . . It appears as though a procedure, once established, seldom dies a natural death; it usually has to be murdered in cold blood if it is to be done away with at all. Probably one reason why procedures are taken for granted and allowed to continue *ad infinitum* is that they were originally set up to eliminate thought.[15]

The technique of bypassing intermediate subordinates both in seeking information and in giving orders will be referred to as the "jump method of direct management." It depended upon the ability of the chief executive to retain sufficient details in his mind to enable him to converse on the technical aspects of

[15] B. B. Gardner and D. G. Moore, *Human Relations in Industry* (Chicago, Richard D. Irwin, Inc., 1950), 195-96.

subordinates' operations. This method of seeking information served several purposes in Blazer's opinion, as expressed in an interview:

It keeps me educated, not only informed. My education in this manner is a refresher course. I pick up information that many times is more significant to me than it would be to the superintendent over the particular operation. . . . Maybe I am more interested in details than the superintendents are and maybe I am drilling them on it. I very deliberately put them on the spot sometimes by asking them questions that I already know about their own plant—that I think they should know—and I know they feel that they won't get caught that way again.

While this jump method of direct management was often used to seek information, it was also used to give orders. Of course, this procedure violated the scalar idea in which orders are passed down to the various levels through the complete line. In order to avoid the obvious disadvantage of undermining the authority of the intermediate subordinate, the method required that each subordinate fully understand that it was his duty to inform his immediate superior of orders received from the top level. Any failure of this upward communication would have caused middle management quickly to get out of touch with actual operations and to feel that its function was no longer important in the process. Blazer recognized that this technique had many pitfalls and was dependent upon a previously established high morale and an idea of cooperation to avoid jealousies. He reasoned in an interview:

The only people that we have had any trouble adjusting to our way of operating have been people who have come in from other organizations. . . . You lose efficiency if you make it [organization] absolutely rigid because a lot of things can't be settled intelligently except by contacts between two people. . . . It is everybody's responsibility to not only do their own job but to make sure that others who are affected by their work are informed of something that they might not otherwise know.

215

Such a management concept depended upon low executive turnover. The executives were able to get the "feel" of operations. They developed an *esprit de corps* which provided an atmosphere of mutual help among members of the management team. Dependent upon these necessary conditions, the jump method yielded certain advantages: the decision of top management resulted in immediate action at the lowest levels, and top executives maintained intimate contact with all members of the organization. The result was an increase in the morale of the organization because members felt that they were more closely associated with the top decision-making process. Top executives had better personal knowledge of the capabilities of those on the lower levels and could act accordingly in executive development.

In Blazer's concepts, the use of committees as a means of maintaining coordination was considered slow, inefficient, and expensive. As a substitute for staff and committee meetings or other multilateral methods, he emphasized the use of bilateral communications. The tools discussed below were adapted to the numerous contacts between two operating executives on the same level. Blazer strove for close cooperation through dependence upon informal organization to eliminate any predetermined obstacle that might interfere.

Growing out of this emphasis upon an informal approach to organization, horizontal communications evolved into a system which was not planned and, moreover, not generally recognized by those in the operating processes. This unplanned system will be referred to as "horizontal automatic coordination." It grew as a result of the pressure placed upon an organization tailored for a small firm but confronted with an unusual increase in organizational problems due to a rapid increase in size.

Sidney Swensrud, former chief executive of Gulf Oil Corporation, expressed a similar idea: "I am also far from clear as to how far the decision-making process can be made sort of 'automatic,' or at least less personal. With some decisions, you have to reach conclusions very quickly because there is a deadline hurrying you along. These, I am sure, are *made* decisions, and they may have to be personal within the grasp of the top executive. With others, you have more time, and it has been my experience that if you let the people who are best equipped to do so throw light on whatever problem is under review the right decision will often emerge more or less automatically out of the process."[16]

Each executive on the lower levels knew his general primary function but often was not sure where his authority and responsibility ended and the next man's picked up. Each was motivated to seek counsel with other executives, directly but informally, in cases in which there was no clear-cut concept of responsibility. This bilateral communication (usually by telephone) between two executives on the same level enabled both to plan a coordinated effort to meet the immediate problem. The basis for discussion often was one or several reports which had originated in each of the departments concerned. These written operating reports had been developed by Blazer to give sufficient information for his own decisions. They originally were communications submitted vertically from the lower levels to the chief executive. As the load on Blazer became greater with little use of functionalization and additional levels to relieve him, the reports began to flow horizontally to other departments for their information. Whereas Blazer gave the ultimate orders when the firm was small, increasingly such orders were issued by lower echelons, not through delegation, but through the necessity which resulted from inability to reach the chief executive. At this stage the reports became more important as a basis of horizontal informal discussion than as a control device of the vertical lines of authority.

The result of the combination of these numerous reports and telephone conversations was that each department put into writing what it planned to do on a certain problem and dis-

[16] Carnegie Institute of Technology, *Fundamental Research in Administration: Horizons and Problems* (Pittsburgh, Carnegie Press, 1953), 74.

tributed the quantitative "proposals" to all departments which could possibly be affected. If there was a part inconsistent with another's action, a telephone call resulted. Blazer observed in an interview: "I think that we have five times as much communication between departments as do other companies." Often when the firm was smaller, these reports were consistent; but as the firm became larger, each department's idea of the actions of other executives became less accurate, and more phone calls resulted. Not until 1954 was this system formalized to some extent by the introduction of a functional production control department.

This system of automatic horizontal coordination satisfied a need in the informal organization after the firm began to grow rapidly. The smaller firm (until 1948) found its simple, flat-line, and semifunctional organization equal to the task. During the transition period in which formal efforts to organize were avoided because of emphasis on flexibility, the high morale of the executives, who had learned the *modus operandi* through long personal associations, enabled the firm to maintain a degree of coordination that would seem to be impossible under such concepts employed in the organization of a large firm. In the smaller organization, Blazer had been able to expand his span of control so that his direct method of management was equal to the task. After the sudden and rapid increase in size between 1948 and 1950, this method of management became more difficult. Even this chief executive with his immense capacity found it more difficult to retain contact directly with all elements in the organization.

Blazer's capacity appeared to result primarily from two factors: the mental ability to remember details and the hours per week devoted to the business. Throughout the history of Ashland Oil, his duties with the company were both vocational and avocational for him.

Although there was no conscious change in the techniques of management, gradually the department heads and the lower echelons of executives evolved this system of automatic horizontal coordination. The final step of recognizing a clear formal organization of the top level failed to develop by 1956, although there were signs of efforts by subordinate executives to formalize their immediate functions.

The development of this informal system resulted because Blazer thought that bilateral methods of communication had basic advantages over the committee type of organization. No committees of any importance operated until 1951. At that time a budget committee, an annual report committee, and a credit committee were the primary examples of multilateral discussions. Actually, Blazer's device of encouraging conferences among two or three members of the executive staff approached the idea of committees. On many occasions after 1950, three or more executives would be involved in telephone conversations that lasted from fifteen minutes to an hour.

After the firm had increased in size, luncheon meetings of the top executives served often as informal committee meetings. A new building finished in 1954 provided an executive dining room which facilitated informal discussions. In this way the team spirit was used and the usual disadvantages of committee actions were avoided. An operating executive commented that this private dining room was used much more than the conference room and was worth its cost aside from its dietary function.

Through bilateral techniques of conferring with his executives, Blazer was able progressively to supplement his factual information, obtained directly from supervisors, with opinions from top executives. When he made a decision, he took account of opinions of all executives involved in the actions. This approach left the final decision to Blazer; it could be said that there was little delegation of decision-making powers. On the other hand, the ideas of the chief executive could easily have changed progressively after each bilateral conversation so that the ultimate decision reflected the impact of the opinions of all who had been contacted. In many cases, the subordinate was not conscious of his importance in the final decision.

This bilateral method depended upon the competence of the central figure. Several important shortcomings of the method were expressed by an employee, trained in another organization, who had not adjusted his thinking to this unconventional type of management: each subordinate failed to get the advantage of the thoughts and reasoning advanced by other subordinates; the subordinate first contacted often found that the ultimate decision was quite different from that which he assumed to be the result of the initial conversations; of greatest importance, the subordinate did not obtain the experience of shouldering responsibility for a final decision.

Concepts of both vertical and horizontal communications within the Ashland Oil organization depended upon the use of efficient mechanical media. Need for frequent exchanges of comments by executives on the same level increased the use of telephone contacts. The informality surrounding the organization encouraged a telephone call in event of any doubt. In this emphasis upon quick changes of policies, the telephone was used in the place of written memoranda in more formal systems. When operations became decentralized geographically, private telephone lines connected important divisions.

Informality and speed were achieved through direct contact of major executives. Each executive had two telephone lines to his office—one for his secretary and one for himself; an employee could call the executive directly without first talking with the secretary. The family atmosphere encouraged employees to contact executives in any event that appeared important. Restrictions on access to executives remained minor even after the attainment of larger size.

While written memoranda were infrequently used, quantitative reports and estimates were circulated widely. Blazer's interest in detail resulted in a large monthly operating report which was distributed to major executives. The section pertinent to each department was given to its department head. This information enabled each supervisor to obtain facts of his own progress for his own control.

Blazer commented in an interview:

The average person does not in my opinion understand and fully appreciate the importance of accurate information and details. Most people, not all of them, have pretty good judgment based on facts as they have them, but they often make wrong decisions because they didn't have enough facts. Many decisions have pros and cons in them and you can find many reasons for handling them this way or that way, but the final decision rests upon where is the preponderance of information. That is the thing that I have kept trying to drill into them and that's one of the reasons that I think justifies my working with a lot of people in that I think that I have a lot of people way down the line trained in my way of doing business.

The informal method of organization depended upon close personal relationships with the lowest employees. Every chance was taken to write to employees at favorable times. For example, one reason for originally paying the "employees' dividend" quarterly was to give the chairman of the board a medium through which he could explain to employees the problems facing the company. Blazer expressed his idea to the board in March, 1948:

I believe the employee-dividend plan will install considerable interest in our stock and in our earnings. It gives us a vehicle to encourage that interest by the sending of a letter each quarter, along with employee-dividends, giving such information concerning current operations and earnings as we believe will be interesting to our employees.

Blazer also considered that contacts between the management and the families of employees yielded high returns in labor relations.

221

Departmentation under Blazer's concepts depended primarily on the individual traits of the personnel and only secondarily on criteria of departmentation that appear logical and rational in an "ideal" set of circumstances. This view is supported and criticized by many in the management literature. Emphasis is placed upon personal factors which cause executives to maintain cooperative efforts. Under this concept it is believed that the establishment of rigid policies of organization distract the executives from the essence of the problem—cooperation and imagination. If responsibilities were outlined completely, the incentive of each individual would be toward a betterment of his own position in the rigid requirements of the hierarchy, whereas the primary effort should be toward more complete cooperation and coordination of the entire group.

The building of an organization under this philosophy is a "natural evolution of responsibilities." This phrase appears in various forms in Blazer's letters, but it became more frequent and more clearly stated in 1954. A comment in an interview in that year will help to clarify this viewpoint:

[In the NRA] we formalized this with the idea of eliminating some wasteful competition but that brought up two other things and then you had to fix that and that brought up 15 other things and then you had to fix them and the first thing you knew the whole structure fell of its own weight—too many regulations. . . . Human nature is too complex and business is too complex—no one can compensate for all of those variables. It can't be done with the human mind—it can't be done by regulations—it's got to be done by competition and evolution. It has to be done with the confidence that what is best will survive and what is not best will sometimes be a source for worry, but in the aggregate, things that are good will succeed and things that are bad will fail. . . . In my organization I come back to that same idea. I put people in jobs that call for a title and a public announcement. I don't give them the title right away. . . . I don't perceive accurately the outlines of that job any more than I can visualize fully the capacity of an individual in a new capacity. . . .

222

Most people think that you can say that logically a thing falls in a certain place. Well, maybe it does and maybe it doesn't.

The idea is that each new executive should be introduced into the going concern with specific duties; he should gradually increase his scope of understanding and fit himself into the company's "way of doing business." This way of doing business was not concretely defined and in most instances involved what Barnard has called "informal organization."[17] The idea was to refrain from setting a fixed pattern which would form a fence for the developing executive. A specific list of duties and responsibilities hinders the aggressive leader from becoming better acquainted with the more general problems.

Blazer used as an illustration one of his vice presidents who was both able and aggressive. He had been one who had pressed for more formal organization. Yet, as Blazer pointed out, he was the best example of a top executive who could not be fitted into a formal chart of authority. Often he would study problems as a staff executive in one of the line departments, only to end up assuming some of the line responsibility. It was claimed that much of his energy and effectiveness would have been lost if he had been completely fenced in in a definite spot. Blazer claimed that the vice president was making a much better top executive because he had had a variety of experiences.

If the specialized job in which he had been placed did not fit his personal capacities, the young executive could gravitate toward that area in which he could contribute the maximum to the welfare of the company. If "positions" were established into which persons were fitted, Blazer felt that the company would be likely to lose the advantage of unique capacities of developing executives. Positions could be changed more easily than personal capacities.

In all organizations there is the "informal" which exists independently of the formal but which can be used to strengthen

[17] Barnard, *The Functions of the Executive*, 114-23.

EXHIBIT 21—EXAMPLES OF EXECUTIVE DEVELOPMENT, ASHLAND OIL & REFINING COMPANY

Executive: Birth Date, Education	Previous Experience	Initial Job with Company	First Executive Position	Intermediate Position	Title of Major Position (1956) and Principal Executive Duties
Everett F. Wells Jan. 18, 1905 Illinois '26 (Economics)	No previous full-time employment	Refinery operator (1926-1928)	Assistant sales manager (1928-1930)	Sales manager (1930-1949)	Executive Vice President (Director)—Top-level coordination of refining, transportation, marketing; relieves chief executive of operating duties
P. C. Talbutt May 25, 1899 Three months of college	Shipfitting; dry cleaning; fireman at oil refinery	Dubbs cracking plant operator (1926-1928)	Assistant refining superintendent (1928-1929)	Division sales manager, Cincinnati (1932-1949)	Vice President (Director)—General sales manager
A. S. Chamberlain July 4, 1900 Yale '22 (Engineering)	Soap company foreman; engineering consultant	Industrial engineering contant (1932-1933)	Personnel manager (1938-1950); safety	Employee magazine; credit union; methods studies	Vice President—Management engineering; pipelines; insurance benefits; economic studies; community projects
E. W. Seaton April 15, 1894 Yale '16 (Engineering)	Iron and mining company; shoe company treasurer; banking	Credit manager (1932)	Credit manager	Treasurer; coordination with Defense Plant Corporation	Treasurer (Director)—Treasurer; real estate; credit manager
Arthur J. Points July 19, 1904 G'town (Ky.) '25 (Liberal arts)	Bank teller	Accounting clerk; two summers in refinery	Accounting (1939-1945); adviser to chief executive	Assistant to president; arbitration; annual reports	Controller—Functional executive in accounting; investigator in all mergers
Edward Emrick Feb. 17, 1909 Illinois '31 (Accounting)	Dairy cattle; dry goods shipping department	Refinery operator (1930)	Assistant treasurer; assistant credit manager	Director of purchases; special assignments	Director of Purchases; Manager of Production-Exploration-Finance
R. A. Whealy Nov. 25, 1907 Iowa '33 (M.S.) (Chem. eng.)	Schoolteacher; research assistant at university	Chemist (1935-1936)	Chief chemist; asphalt sales	Assistant (1943), superintendent of refineries (1946-1953)	Coordinator, Refining and Sales—Executive assistant to executive vice president

the formal organization. If the formal organization were preconceived in a rigid manner involving precise definition of duties, Blazer believed that the chances were that the informal organization would be inconsistent with the formal and would weaken rather than strengthen the formal structure. His concept of "evolution of responsibilities" referred to an effort to allow the informal pattern to become evident prior to conscious decisions relating to formal organization. Often this "natural" construction might not have appeared as logical as one would have liked, but it had the advantage of being effective in gaining the maximum cooperative effort in those cases in which it was possible to develop slowly human relationships in the organization.

Under this natural evolution of responsibilities, Blazer preferred not to make a promotion until the individual executive was allowed gradually to obtain experience, often as a "staff assistant." The flexibility of job titles made it possible for an executive to work on a job in an "acting" capacity prior to recognition of the change. If the executive proved that his abilities and interest qualified him for promotion, an announcement ultimately was made. Often the new title was not the same as the one held by his predecessor; the job was tailored to the person's abilities. Exhibit 20, therefore, can only give a first approximation of the personnel that performed the various functions. If the executive proved that he was not able to fill the duties of the new job, Blazer could easily shift him to one more suited to his abilities and yet avoid loss of status for him.

Ashland Oil could be continually in a state of reorganization as a result of changing conditions without the confusion that usually develops in periodically changing the formal structure. At the time of each of the mergers, this evolutionary process enabled Blazer quietly to work out the new management without sudden changes that could adversely affect the morale of the workers. In the case of Allied, the entire operation remained separate with the exception of refining, but gradually the home office absorbed certain overhead personnel. In the case of Aetna, the individual around whom the employees identified themselves remained as a symbol of a continuation of past policies. There was no sudden change in the management. The problem at Freedom-Valvoline was more difficult. Three executives were potential material for the chief position. The actual change was made only after a year, during which the division was operated by a committee of the three. In the case of Frontier, one of the owners remained temporarily to give the new management a chance to become established. An executive from Allied spent part time in Buffalo and gradually evolved into the position of executive vice president of Frontier. In each of the cases, time was taken to secure the effects of the informal relationships prior to establishing the formal position. The pliable characteristics of the system permitted changes to occur without attracting unfavorable attention of employees. The fact that the status of an executive was dependent upon his actual functional relationships in place of scalar or hierarchical positions made the growth of the organization correspond closely with current necessary functions rather than with preconceived and static patterns of an engineered structure.

Blazer's idea of evolution of responsibilities in organization through the adaptation of the organizational structure to the individual executives resulted in informal executive development procedures. From the very beginning of the company, the training of executives was a primary problem. Consistent with his basic philosophy, Blazer did not set up an executive development program in the sense that the phrase has been used in formal plans. Nevertheless, the stress he placed upon the general evolution of each potential executive contributed to a solution of his needs for capable executives in the rapidly growing company.

The major executives in 1956, shown in Exhibit 20, came primarily from two sources: young executives hired by Blazer in the early 1930's, and key executives of merged companies who remained in the combined organization after the mergers. Exhibit 21 is a summary of the principal duties of seven of the

226

executives who had been with Ashland Oil for twenty or more years and who in 1956 were major executives.

The philosophy of management discussed in this chapter gives great weight to flexibility and informality. The concepts were important in the early success of the company; they had characteristics that contributed to its rapid growth; but they posed a number of problems as the firm became large.

10

SUCCESS AND SUCCESSION

The Evaluation of Policies and Practices

THE PROCESS of evaluation is the measurement against certain standards considered to be valid by the appraiser. In the theory of management there are many proposals for such standards. Most depend upon their logical appeal and their mutual consistency for their proof. Concepts are said to be "good" when they agree with these defined standards, and "bad" when they differ. At their present stage of development, these standards are inadequate as a basis for evaluation. Often they result in ambiguous and contradictory recommendations. The student of management, therefore, must use an evaluation procedure that measures in terms of objective results.

Intensive case studies provided a fruitful approach for research in administration. In the case of Ashland Oil it has been obvious that repeated successes have been achieved by Blazer over a 33-year period. As the number of observed successes increased, the probability that chance was the cause has decreased. The writer has studied the numerous situations of decision making in search for the basic causes of this record of successes. Now it is time to point out the generalizations that logically follow from this historical study of administration in action.

Several factors that more or less were out of the control of the management were necessary but by themselves not sufficient for the 33-year record. The growth in the demand for oil products provided the opportunity for good managers to direct profitable ventures. The social policy of the federal government in its control of monopoly practices was of value to the operation of smaller companies. The geographical characteristics of Ashland's area of operations provided important advantages for a company that was located there. Yet these reasons were generally applicable to other companies which failed. A look into internal administrative processes is required to find the sufficient causes.

Two groups of factors within the control of the management answer the question as to why the company was successful. First, the operating policies have set the guides to proper decisions. Second, Blazer's managerial concepts have implemented the direction of the company in execution of the policies. The results have been due to exceptional abilities and ideas of a single chief executive. In this case, a single executive has provided the top-level direction over the entire period. The history of the company has become a specialized biography. This study of the policies and concepts employed by him contributes valuable empirical data to management research.

The most important of the general operating policies was continual emphasis on the interstices of the oil industry. Blazer searched for and found niches in all four branches of the industry. He considered that he was not as interested in the conventional approaches to operations as he was in the advantages of unconventional actions. For example, he was more afraid of governmental help for small business by guaranteeing a fixed place in a rigid pattern that would prevent expansion into new areas than he was of the advantages of scale held by the larger companies.

Preceding chapters have offered instances of the application of this policy of filling niches. In the production branch, it was seen that Ashland Oil obtained its greatest success in the operation of "stripper" or marginal wells. Executive attention was

229

directed to decreasing the cost of operations of these producing wells which others found uneconomical to operate. Most of the remaining company requirements of crude oil were bought from independent producers through its large purchasing organization. Only at times of exceptionally favorable tax treatment of expenditures in exploration and development did the company follow the view held by many executives in the industry and expand its company-owned crude oil supply.

The management placed great importance on river transportation as the niche in which it was able to obtain an economic advantage. This specialization on river barges provided a means by which the company was able to meet the competition of pipelines operated by larger companies. Even when Ashland Oil acquired a pipeline—the Cumberland—it did so at a time when no other company believed that operation of the facilities would be profitable.

In the refining branch the management acquired obsolete equipment of little value to anyone else and continued to operate it long past the time when other companies considered its use desirable. Even though the management shifted to the use of new refining techniques when absolutely necessary, it retained the obsolete equipment and continued to operate it profitably. Whereas other refiners emphasized automation and volume production, Ashland Oil, as a small company, effectively exploited the niche of production of low-volume, high-margin specialty products.

In the marketing of refined products, the management filled the geographical niches left by major companies. It specialized in offering service along the economic ridge of costs between the eastern seaboard and the midcontinent oilfields. Whereas the managements of many large companies regarded an increase in their market share in a given area as a criterion of success, Ashland's management tried to remain an insignificant marketer in any given area and to expand by moving into other profitable areas. It specialized in the sale of products in areas of low-sales concentration which were relatively unattractive to the larger companies.

As a small company, this specialization in the niches of the industry proved to be an important reason for success. As the firm grew, it became increasingly difficult to find niches sufficiently large to accommodate the company's complete operation. Growth forced a greater emphasis on conventional phases of operations. The company sought intermittently to expand its exploration for crude oil; it found that participation in pipelines was desirable; it promoted branded gasoline. As the company grew, its operating policy remained one of emphasis on interstices but necessarily became more like those of large firms.

A second general operating policy was the continual adjustment of its pattern of integration. Previous discussion has shown that repeated changes in Blazer's opinions were characteristic throughout the company's history. These resulted partially from financial inability to maintain the rate of increase of company-owned crude oil production and company-owned marketing outlets equal to the rate of increase of refining capacity. Each step in the growth of refining brought with it additional pressures to acquire greater security in the supply of raw materials and marketing outlets. These pressures caused Blazer to change his reasoning quickly in the planning of capital expenditures. His opinion that independent producers and jobbers could perform their function more efficiently than could his company-owned departments continually reappeared as conflicting with the fact that supply by independent companies was more insecure. The result was that he made repeated changes in his rationalizations to the board. Although the fluctuation of capital expenditures among the four branches fitted the dynamic situation in the oil industry, the management probably failed to achieve greater returns from investments by its lack of a planned capital expenditure program in production.

The reluctance of the management to set any fixed integration pattern had an important impact on the nature of growth of the company. Since operations were based upon the assumption that throughput should always be kept near capacity, refining became the reference branch. Other branches were forced to balance their operations with refining throughput. In times of decreased demand for oil products, the sales branch became a potential limiting factor to economical refining operations. Although the area of competition was extended in order to maintain volume, it was often necessary aggressively to seek new independent jobbers. If sufficient outlets were not available or if other companies purchased Ashland's jobber outlets, the management became especially receptive to offers of merger from independent companies. During these times of decreased demand, most independents were experiencing financial problems. As a result, both parties looked favorably on mergers.

Previous discussion has shown that in most of the mergers, marketing facilities were the chief interest of the Ashland Oil management. The two periods in which most mergers took place were ones of decreased business activity in the oil industry (1930-1931 and 1949-1950). Purchase of these marketing properties facilitated adjustments in the integration pattern necessary to keep refining throughput at capacity but was not the result of a formally planned program of expansion. This policy of giving primary importance to refining thus led to growth in marketing at a time in which properties could be obtained on a most favorable basis.

Although the majority of the mergers were with independent companies which possessed good marketing outlets, the fact that each company also owned a refinery provided the foundation for still further growth. In every case the refining equipment was small, obsolete, and not valued highly in the negotiations. Yet after the mergers, the Ashland Oil management quickly expanded throughput to exceed that attained by the former company. In this way the initial attempt to seek marketing outlets resulted in still greater increase in throughput, and in turn, new pressures for sales outlets.

232

The result of this opportunistic policy in integration was that the company acquired marketing facilities and refining capacity at a reasonable cost. It grew rapidly in spite of the lack of a strong desire on the part of management to expand. Successful negotiations and profitable operations made growth inevitable.

Ashland's policy on integration led to a continued use of a third operating policy that had originated at the very beginning of the company. The Ashland Oil management recognized the sporadic nature of refining improvements in the industry and the indivisibility of additions to equipment. Consequently, as a result of limited financial resources, Blazer decided to refrain from adding new equipment until it became clear that an innovation would affect the entire industry. Only after unusually rapid changes in technology was he willing to make improvements which had forced many independents into poor financial condition. Once it was clear that the change would be necessary, he added equipment of sufficient size in order to obtain reasonable economies of scale. In most cases, this increased capacity reinforced the pressure on the sales branch, since the company could not expect to increase outlets for refined products as quickly as it increased refining capacity. The incremental cost of the extra capacity was so low that the management preferred to extend its marketing territory and accept lower netbacks in an effort to attempt to increase volume rather than to depend upon the security of a smaller but more stable throughput. This policy was a logical one for a small company that is operated with a flexible netback and a management that was not satisfied to accept a static sales volume for an extended period.

These three general operating policies help explain the manner in which the small company operated profitably in competition with very large firms. The execution of these policies further explains the company's growth. Blazer developed his own concepts of administrative behavior that achieved especially good coordination.

The concepts of technological, marketing, financial, personnel, and organizational flexibility have provided the keystone of Blazer's managerial philosophy. The evidence shows that these ideas are interrelated and especially suited to growth. They have resulted in high morale of executives and employees, while at the same time their successful application has depended on the existence of high morale. The problem that confronts the student of management becomes one of reconciling these ideas with current, apparently conflicting, thoughts—or he must discard one group. Some reconciliation can be achieved. The discarding of Blazer's thoughts would be unwise in the light of the evidence as to their success under the situations faced over the thirty-three years. Additional research on the success of such ideas in other companies can provide a more conclusive answer. This research should be focused on the smaller growing companies, inasmuch as the likelihood of finding such philosophies will be greater in those companies than in the large companies whose managements have been molded in the pattern of the current thought of the "professional manager."

What useful generalization of administrative behavior is supported by the evidence submitted in the preceding chapters? It is evident that the degree of flexibility has been a most important factor in the growth and profitability of Ashland Oil. Some degree of flexibility is generally agreed to be desirable. The question of the correct degree should be answered in the light of returns or advantages from each type of flexibility balanced against the costs or disadvantages of each.

As Ashland Oil grew rapidly after 1948, it appeared that the returns from the degree of organizational flexibility achieved up to that time were diminishing and the costs of maintaining such a degree were increasing. The most important factor that caused the net returns to decline was the increase in the scale of operations. Other factors that could have been important in explaining the increase in costs to this company, for example, changes in governmental policies and changes in personnel in the company, were relatively constant between 1948 and 1957.

The evidence that a high degree of all types of flexibility contributed to the growth of the company is convincing. This same growth caused the costs or disadvantages of this degree of flexibility to increase as the returns decreased. It therefore follows that in the continued growth of the company a point has been or will be reached where the management should emphasize a lesser degree of flexibility. Up to this point the emphasis yields increasing net returns; after this point the returns are less.

This relationship of changes in scale to net returns from a given degree of flexibility appears to be valid for each type discussed in Chapter 9. The point of maximum net returns, however, tends to differ for each type. Since each tends to reinforce the advantages of the other types, the determination of the overall point of maximization is a complex problem. Furthermore, these points of maximum advantage are functions not only of the scale of operations but also of the rate of growth, the nature of the industry in which the company operates, and the personal characteristics of the executives.

In the past discussion it has been shown that the structure of the oil industry, especially its composition of very large companies and a number of much smaller ones, has affected Blazer's orientation to his problems. The industry's low ratio of labor costs to capital investment has enabled the company with larger assets and sales to employ those administrative techniques characteristic of a company with a small total executive and labor force. Furthermore, in the early period (1924-1933) the instability of prices and production in the industry demanded sensitive communication and organization patterns that would enable quick adjustments to the changing situation.

In addition to scale of operations and nature of the industry, the degree of flexibility that was desirable has been determined by the personal characteristics of the company's personnel. The leadership of Ashland Oil has been provided by a man who built the organization from the beginning and developed a charismatic atmosphere in his relationships with his subordinates. The subordinate executives were young, inexperienced, and potentially malleable to the uniquely functioning organization. The subsequent low turnover of personnel made it possible for each executive to become thoroughly familiar with the *modus operandi* of the organization. The confidence of the stockholders in the chief executive permitted him to direct activities as if the company were owner-operated. The "Ashland family" idea and its resultant high morale was engendered by the location of the company's major facilities in a small town.

The point of maximum net returns from a given degree of technological flexibility was early affected by the extent of use of automation in the industry. After the change from the batch process to the continuous process in the late 1920's, most larger companies stressed economy of volume production rather than flexibility of products. This action by the other companies left a place for Blazer to maximize returns by applying his idea of technological flexibility. His use until 1938 of shell stills in series with fractionating towers is a good example of his success in the maintenance of returns in this type of flexibility above its costs for a longer period than was characteristic of the larger companies. After its own rapid growth in 1948-1950, Ashland Oil, as a larger company, continued to emphasize volume production, which had been practiced during the World War II years. This policy, which resulted from the increase in scale, caused the management, in the period of depressed prices in 1954, to recognize that its ability to adjust to new products had become smaller. Volume throughput had impaired technological flexibility.

The advisability of emphasis on a high degree of personnel flexibility is clearly a function of the scale of operations. As Ashland Oil grew, the "family" became so large that it was impractical for all executives to have personal contact with all others. Geographical dispersion increased the problem. Yet the continuation of the application of this concept was successful even in a company of 4,600 employees. The characteristics of the personnel discussed above were important in this successful application in such a large company. Moreover, the experience and proficiency of all executives for handling their decisions on an informal basis increased the desirability of this concept. In addition, the number of hours per week that each executive devoted to company duties remained exceptionally high and permitted greater use of both the "generalist aspect" and the "personal aspect" of personnel flexibility. It is evident that one group of managers will be able to utilize successfully the ideas on a given scale where another group of equally competent executives may not. Ashland's management represents a case of continued successful application in a relatively large firm in spite of the increased costs resulting from growth.

Flexibility in marketing has remained valuable throughout the firm's growth for several reasons. First, operations in the "territories" of six different reference marketers has required that the firm's marketing policies be especially adaptable to the actions of its competitors. Second, the firm's rate of growth has required such a rapid increase in marketing outlets that multiple channels and use of independent jobbers have remained financially necessary. Third, in spite of the rapid absolute growth of the firm, the company's relative importance in most of the territories remained small. It appears that the returns from marketing flexibility were still relatively great for the firm's size in 1957 as compared with the cost of such flexibility; however, the critical size requiring a different approach was only slightly larger.

One aspect of financial flexibility tended to become more potent as the firm grew, while another generally decreased in usefulness. The shiftability of assets became more difficult as the firm committed itself to greater fixed assets in the different branches of the industry. On the other hand, after 1945 the ability to secure additional funds in the securities market enabled Ashland's management to take advantage of merger opportunities which occurred on short notice. The net effect was that as a larger firm, Ashland Oil attained a different kind of financial flexibility than it had as a small firm.

The five flexibility concepts give content to an important explanation of Blazer's success. His "bargaining ability" was founded upon these ideas. In the 1930's he was able to sell the capacity throughput of his refinery during the Great Depression through application of these ideas. In the late 1940's, because of emphasis on these fundamentals, he demonstrated his ability to act quickly when an opportunity developed. Without these managerial concepts, it seems unlikely that he would have been able to adjust so skillfully to the dynamic situation.

While the gross advantages of these concepts of management have been necessary to the management of the small and rapidly growing company, their potential disadvantages, that is, their costs, have been significant. By far the greatest of these disadvantages is the centralization of key elements of decision making in the chief executive. Although special attention has been given to the development of executives of general ability, the acceptance of specific responsibility for a decision on the part of subexecutives was not clear. In fact, in 1954 the company had many "good general vice presidents" but might have found that the succession problem was critical. The problem, however, arose not from the inability or inexperience of the subexecutives in the general functions of the firm, but from the psychological problem on the part of a potential chief executive to develop confidence in his own final decisions.

The succession problem became not one of the generally discussed maneuvering on the part of several potential chief executives, but one of developing confidence on the part of an executive to fill the "old man's" shoes. In the case of Ashland Oil, several executives had qualities that could have provided good management for the future; however, their problem of convincing themselves, the board of directors, and the public that they could successfully manage the company remained. The full impact of this problem never was felt, since nothing happened to shift suddenly the responsibilities from Blazer to someone else. This type of managerial philosophy caused Blazer to retain complete control over a long period of time when the risks of loss of his services were great.

It was fortunate that during the last three years of control by Paul Blazer (1954-1957) greater responsibility was given to the subordinates. Gradually, Blazer began to test how the company could run without him. On one occasion he went on a Mediterranean cruise, when he was more out of touch with immediate problems than he had ever been previously. Before, the mere fact that he was in Florida did not mean that he was not actively running the company, for daily long-distance calls were used to keep in touch with the home office—no different from his calls from his home while he was in Ashland.

The idea of the chief executive function being handled by two men became attractive for the large company as it shifted from the administration of Blazer. The techniques employed by Robert E. Wilson and F. O. Peake of Standard Oil Company (Indiana) had been noticed by the Ashland Oil management. Moreover, President Rexford Blazer and Executive Vice President Everett Wells were uniquely qualified for this type of management.

Rex Blazer had operated for twenty years in Allied, using managerial techniques different from those of his uncle Paul Blazer. He had always had a personality which fitted sales and public relations. He preferred to delegate more authority and responsibility than Paul Blazer had. He was very active in civic organizations and trade associations. In short, he was ideal for the "outside-directed" man for the top management team.

Everett Wells, on the other hand, had worked for Ashland Oil since his graduation from college in 1926. His college training in business gave him the background to go into sales. He gradually had been given more authority, and during the last ten years he had been the chief coordinating officer among sales, transportation, and refining. He had worked so closely with Paul

Blazer that his managerial ideas were similar. In short, his knowledge of operations of the company made him ideal for the "inside-directed" man for the top management team.

Blazer's complete philosophy yields a paradox. The philosophy of flexibility had been so important that the chief executive had actually become inflexible in his own administrative behavior. He had realized that his techniques had been of greatest value in a smaller company, but he had been unable to change his techniques to fit more nearly the needs of the larger organization. Since he had become so proficient in the application of his flexibility concepts, his philosophy of management, so exceptional in the direction of the small firm, remained the best technique *for him* in the larger firm. Even for the chief executive who continued to fight the trend toward greater formalization of authority, the use of formal types of organization evolved. Exhibit 20 has graphically shown this evolution from simple line organization, through the addition of staff assistants, to the use of functional specialists, and finally, to some decentralization by product division.

The temperament of certain individuals is such that they seek decision-making powers. It is no accident that many of the notable successes in the promotion of a business have been the type that would emphasize centralization of decision making. A man who has built a company primarily through his own energy and skill finds that it is most difficult to adapt his technique to a situation that demands decentralization and delegation. The executive who has always made decisions by initiation finds it very difficult to make the majority of his decisions by approval. Yet the time comes that a firm reaches the critical size which requires a different type of administration. The fact that new tools are needed for administration in the larger firm does not offer evidence *per se* that these new tools should have been used throughout the firm's development. Nevertheless, the problem of this inflexible emphasis on flexibility potentially causes a major disadvantage when costs of flexibility exceed returns.

240

In conclusion, this study has shown the value of a group of policies and concepts to the success of a company. Although the policies are not in agreement with those of other leaders in the oil industry and although the concepts conflict with certain "principles of management," they obviously have been useful to this management. In fact, the chief advantage, in the writer's opinion, that this management has maintained throughout its history has been administrative creativeness unfettered by the shackles of a "cookbook" solution to its business problems. It is this ingenuity to seek techniques uniquely suited to the specific problems that is the principal administrative strength illustrated. Further research should be directed toward intensive case studies of other situations in search for new ideas to improve the art of management.

INDEX

Abbott, Walter, 105, 106
A. D. Hambleton & Company, 30
Advertising program, 161
Aetna gasoline, 159, 167
Aetna-Louisville, 141
Aetna Oil Company, merger of with Ashland Oil, 2, 104-106, 131, 211, 225-26; history of, 105-106; mentioned, 103
Aetna Oil Service, Inc., 105
Albin, A. C., 28-30
Aledo, Ill., 5
Alexander, C. M., 105, 208-209
Alexander, L., 208-209
Allied-Ashland, 141
Allied-Ashland Tankers, Inc., 98, 130
Allied Oil Company, merger of with Ashland Oil, 2, 95-101, 130-31, 211, 225; history of, 96; mentioned, 147, 177
Alteration, techniques of, 43
American Independent Oil Company, 88-89, 98, 133
American Institute of Mining and Metallurgical Engineers, 152, 182
American Mineral Spirits Company, 147
Amurex Oil Development Company, 133
Armour, Ogden, 24
Arnold, H. C., 208-209
Ashland, 127
"Ashland," used as brand name, 156, 157, 158, 166-67
"Ashland family," 57, 183-84, 207, 236. *See also* Family concept
Ashland gasoline, 166-67
Ashland Guatemala Company, 89
Ashland Oil & Refining Company, early history and growth of, 1-3; area of operations of, 15-21; formation of, 40-41, 68-69; and specialty products, 56, 147, 165-66, 230; and concept

Ashland Oil *(continued)*
of flexibility, 61, 82, 92, 101, 190, 192-204, 233-38; and exploration for oil, 69-71, 86-92, 132-36, 231; and modernization and improvements in refining, 71-76, 136-39; organizational structure of, 76-77, 202-13; postwar transition and adjustments of, 77-89, 101-104; expansion of through mergers, 93-116, 130-31, 225-26; and use of river transportation, 115, 117, 120-30, 141-42; transportation system of, 115, 117-44; and use of pipelines, 115, 121, 124, 129, 130-31, 139-44; pricing policies of, 146-47, 169; and marketing, 151-69, 195-98; and promotion of brand names, 156-61; market share of, 161-63; and Blazer's concepts of role of chief executive, 170-89; and Blazer's objectives of management, 181-87; under successors to Blazer, 187-89, 239-40; internal communications of, 213-21; and evolution of responsibilities, 222-27; evaluation of Blazer's management of, 228-41. *See also* Ashland Refining Company, Swiss Oil Corporation
Ashland Oil & Transportation Company, 65
Ashland refinery. *See* Catlettsburg refinery, Leach refinery
Ashland Refining Company, organization of, 9; Blazer's management of, 33, 41-69 *passim*; consolidation of with Swiss Oil, 40-41, 68-69; and Leach refinery, 47-50; and offers to purchase, 58-59; during depression of 1930's, 61-63; and purchase of Cumberland pipeline, 63-65; effect of government policies on, 66-67. *See also* Ashland Oil & Refining Company

INDEX

INDEX

Organization *(continued)*
bility in, 202-204, 210, 212, 234; Blazer's concepts of, 204-27
Otis and Company, 30
Out-of-pocket costs, 85
Owens, Erskine, 188, 208-209
Owensboro-Ashland Company, 124, 132
Owensboro Pipe Line, 124-25

Paintsville, Ky., 28
Paul Blazer, 127
Peake, F. O., 239
Pennzoil, 160
Personal relationships, of Blazer with oil executives, 59; between personnel and management, 202, 221
Personnel, and continuous operation, 44; of Ashland Refining Company, 50; and development of the "Ashland family," 57, 183-84; and expansion of Ashland Oil, 76-77; opposing views of, 171; flexibility in, 200-202, 235-37; relationship of with management, 202, 221; and Ashland Oil's organizational structure, 204-13; and the evolution of responsibilities, 222-26. *See also* "Ashland family," Family concept, Labor
Petrochemicals, 11, 45, 165
Petroleum, characteristics of, 14-15; marketing of, 18-20; and transportation, 19, 118. *See also* Crude oil, Marketing, Oilfields, Petroleum industry, Pipelines, Refining, Transportation
Petroleum coke, 55
Petroleum Administration for War, 73, 77
Petroleum Exploration Inc., 9, 58
Petroleum industry, beginnings of, 1; Blazer's early contacts with, 7; general characteristics of, 9-13; problems of in 1920's, 27; postwar recession in, 101-104; and weak market, 164; structure of, 235. *See also* Independent oil companies, Major oil companies, and names of individual oil companies and refineries
Petroleum Marketing Divorcement, Congressional Committee on, 153

Phillips Petroleum, 12
Pipelines, construction of, 19, 129-30, 133, 139-41; economic value of, 45, 118-19, 120; common carrier, 47, 108, 115, 142-44; "Little and Big Inch," 81; of Allied Oil, 97; purchased by Ashland Oil, 124, 130, 131; mentioned, 79, 83, 114, 167, 231. *See also* Cumberland pipeline
Pipe stills, 45, 193
Points, Arthur J., vii, 176, 208-209, 224
Policies, for refining, 46, 50-57, 230-33; of pricing, 146-47, 169; for marketing, 152, 154, 166-67; oral communication of, 214; evaluation of, 228-41. *See also* Crude oil, Flexibility, Interstices, Personnel, Production
Polymerization, 43
Price marketers, 147, 197
Prices, of crude oil, 22, 25, 27, 35, 37, 56, 82-85, 102, 103-104, 179-80; fixing of by government, 67; and prorationing, 83; of refinery products, 103-104; Ashland Oil's policies on, 146-47, 169
Pricing, 146-47, 169, 195-98 *passim*
Processes. *See* Batch process, Cracking, Continuous process
Production, of oil in Kentucky, 22-27; in Big Sinking field, 25, 27; by Swiss Oil, 35-37; by Ashland Oil, 82-89, 134; and operation of stripper wells, 85-86, 229-30; retrenchment in by Ashland Oil, 91
—gross, definition of, 84; of Ashland Oil, 87-88; of Allied Oil, 97
—net, definition of, 84; of Swiss Oil, 35; of Ashland Oil, 87-88; of Sohio, 90; of Allied Oil, 97
Products, derived from refining, 15, 43-46 *passim*; specialty, 46, 56, 145, 147, 165-66, 230; brand names for, 53, 156-60 *passim*
Profits, from refining operations, 3, 49-50, 57, 61, 154; and gasoline yield, 51-53; as objective of management, 183
Profit-sharing plan, 180, 184

INDEX

Prorationing, 83
Pryse refinery, 63, 79-80
Public relations, 183
Pure Oil Company, 140, 162
Purification, 43
Pynchon & Company, 30, 31

Quaker Oil Company, 25
Quaker State oil, 160

Railroads, 118, 120, 121, 131
Ralston, Roy, 134, 208-209
Red Pepper Ethyl gasoline, 53, 149
Reference marketers, 19-20, 237
Refineries. *See* Buffalo, Canton, Catlettsburg, Findlay, Freedom, Leach, Niles, Pryse, Refining, St. Elmo, Tri-State
Refining, profits from, 3, 49-50, 57, 61, 154; as part of oil industry, 11; products resulting from, 14-15, *see also* Products; characteristics of, 42-47; Ashland Oil's capacity for, 49, 51, 57, 60, 61, 71, 72, 75, 97, 104, 176-78; Blazer's management policies concerning, 51-57, 230-33; Ashland Oil's postwar transition in, 77-82; postwar recession in, 101-104; flexibility in, 193-96. *See also* Cracking, Equipment, and names of individual refineries
Reports, 217, 220-21
Repressuring with gas, 33
Residual fuel oil, 19, 96, 147, 177, 194
Responsibilities, evolution of, 222-26
Retrenchment, 91, 133, 135
Reynolds, Arthur, 24
Reynolds, George M., 24
Rion, Stokely, 122, 208-209
River transportation. *See* Transportation
R. J. Brown Company, merger of with Ashland Oil, 147, 165, 166
Road oil, 55
Robinson, E. A. G., 191
Rockefeller, John D., 64
Rockefeller Foundation, 64
Rubber, 11
Ruth Ann, 122
Ryan, Paul, 113

St. Elmo refinery, 114, 138
Sales, of Ashland Oil per year, 21; drop in caused by depression, 61; to other oil companies, 151-52; Ashland Oil's policies on, 152; and promotion of brand name products, 156-57, 159-60. *See also* Marketing
Scout, 122
Seaton, E. W., 189, 208-209, 224
Secondary recovery, 33, 85, 135
Seep Purchasing Agency, 37
Senator Combs, 123, 124, 125, 142, 151
Service stations, 11, 104, 146, 147, 152, 157, 158
Shaffer Oil and Gas Company, 30
Shatford, J. E., 7
Shell Oil Company, 151, 152, 162
Shell stills, 51, 71, 193, 236
Shell-Union, 58
Signal Oil Company, 107
Sinclair Refining Company, 39, 140, 162
Sisler, G., 208-209
Sistersville, W. Va., 9
Socony Mobile plant, 179
Socony-Vacuum Oil Company, 19, 108, 152
Sohio. *See* Standard Oil Company (Ohio)
Solvents, 147, 165
Southern Pipe Line Company, 130
South Penn Oil Company, 37
Specialty products. *See* Products
Spindletop gusher, 1
Spraberry project, 132-34
Staff assistants, 203
Staff specialists, 76-77
Standard Oil Company (original), 1, 12, 64
Standard Oil Company (Indiana), 19, 239
Standard Oil Company (New Jersey), 19, 55, 61
Standard Oil Company (Ohio), relations of with Ashland Oil, 59, 151, 152; exploration activities of, 89-90; and purchase of pipeline, 124; market share of, 162, 163; mentioned, 19, 58, 87, 141

INDEX

Standard Oil of Kentucky, 19, 162
Status systems, 210
Stills, 53. *See also* Pipe stills, Shell stills
Stock, of Swiss Oil, 30, 32, 39-40; of Ashland Oil & Refining Company, 69, 198-99
Storage tanks, 144, 146
Stripper wells, 85-86, 229-30
Successors, to Blazer, 187-89, 238-39. *See also* Blazer, Rexford; Wells, Everett
Sumpter, M. C., 208-209
Suez oil situation, 175
Sun Oil Company, 12, 162
Superior Oil Company, 87
Swiss Drilling Company, 23
Swiss Oil Corporation, hires Blazer, 9, 33; early history of, 23-28; board of directors of, 24, 174; revitalization of, 28-33; integration of into refining, 33-40 *passim*, 42, 60-61; merger of with Ashland Refining Company, 40-41, 68-69; and purchase of Leach refinery, 47-50 *passim*. *See also* Ashland Refining Company

Talbutt, Palmer, 188, 208-209, 224
Tank-car marketing, 147
Tank cars, 73, 118, 131, 146-47
Tanker fleets, 117, 131
Tankers, lake, 96-97, 102, 115, 131; ocean, 98, 120, 130-31
Tanks, storage, 144, 146
Tank trucks, 131
Tank-wagon marketing, 147
Taxes, 98, 99, 109. *See also* Excess profits tax
Technological flexibility, 15, 193-95, 236-37
Tecumseh pipeline, 140-41
Terminals, 128, 155
Territories, marketing. *See* Marketing
Tetraethyl lead, 61
Texaco, 162
Texas Company, 37, 55, 58, 79, 110
Texas oilfields. *See* Oilfields
Texas Railroad Commission, 134
Tex Harvey pipeline system, 133

Thermal cracking. *See* Cracking
Tick Fork, 22
Tide Water Associated Oil Company, 137
Titles, for oil leases, 23; of executives, 207-13
Titusville, Pa., 1
Top leases, 28-29
Torrent, Ky., 24
Towboats, 84, 96, 117, 122-30, 141-42, 194
Towl, Forrest M., 64
Transportation, costs of, 18, 65, 81, 146, 151-54 *passim*; use of river for, 18-19, 49, 73, 96-97, 120-31, 141-42, 230; by horses, 24; by tank cars, 73, 118, 131, 146-47; equipment for of Ashland Oil, 115, 122-32; economics of, 117-19; by pipelines, 119, 121, 124, 129-30, 139-44; flexibility in, 121, 193, 194; importance of in Ashland Oil's success, 144. *See also* Barges, Pipelines, Railroads, Tank cars, Tankers, Tank trucks, Towboats
Tri-State, 127
Tri-State Aviation gasoline, 149
Tri-State Ethyl gasoline, 149
Tri-State Refining Company, merger of with Ashland Refining Company, 60-61, 65, 149, 151, 166
Tri-State Super Motor gasoline, 149
Tube-and-tank process, 55
Tulsa, Okla., 70
"Turnaround," 44

Union Gas and Oil Company, 28-33
Unions, labor, 66, 184, 202
United Oil Company, 23
Universal Oil Products Company, 55
University of Chicago, 5

Vacuum Oil Company, 58
Valvoline, 141
Valvoline Oil Company, 107. *See also* Freedom-Valvoline Oil Company
Valvoline oils, 107, 159, 160
Van Buren, E. E., 208-209
Vandeveer, W. W., 95, 96, 98
Van Hooser, J. D., 30

252

INDEX

Vertical integration. *See* Integration
Vinson, B. F., 208-209
Viscosity, 96

Wafra field, 88
Wallace, Earl, vii
Waples, W. A., 208-209
Ware, F., 208-209
Ware, Lewis, 208-209
Warren County area, 7, 24
Washburn, J. L., 24
Water flooding, 85, 135
Weaver, Earl, 180, 208-209
Weber, Max, 190
Wells, Everett, quoted on transportation, 140; becomes sales manager of Ashland Oil, 148; quoted on specialty products, 165-66; as president of Ashland Oil, 181, 187-88; execu-

Wells, Everett (*continued*)
tive qualifications of, 239-40; mentioned, 176, 208-209, 224
West, Cecil, 208-209
Western Reserve Refining Company, 79
Whealy, R. A., 208-209, 224
White Pepper gasoline, 149
White Rose gasoline, 112, 159
Wildcatting, 22, 86
William and Vashti College, 5
Williams, Buck, 208-209
Williams, John Fred, 188, 208-209
Wilson, Robert E., 239
Winchester, 142
Wolbert, George S., 143
World War II, effect of on Ashland Oil, 72, 73, 75, 80, 81, 184; and marketing of gasoline, 155-56, 158
Yancey, Bob, 188, 208-209

253